Creating the worlds of

# STAR WARS

## 365 DAYS

JOHN KNOLL

with J. W. RINZLER

Harry N. Abrams, Inc., Publishers

# Contents

# Foreword

I've always thought that the first group of people at Industrial Light & Magic, those who helped make *Star Wars* back in 1976, was really made up of three groups. I belonged to the smallest group, which had come up doing the *King Kong* (1933) style of stop-motion animation with puppets. This group usually didn't have a big budget, so we put all our energy into lighting and setting up a shot just right. We couldn't do anything other than just rent the camera gear and shoot it. The second group at ILM had learned from Doug Trumbull and Con Pederson, via Stanley Kubrick, in England on *2001: A Space Odyssey* (1968). Headed by John Dykstra, they almost thought backward. They would start with, "Hey, here's a shot we want to do that's really neat"—but it didn't end there. If they needed to build a new camera for that sequence, they built a camera; if they needed to build a rig to move a model along, they'd build the rig. That was unheard of in my group. The third group was a lot of guys and gals straight out of college who had never made a film before.

So ILM was a mix of an old-school effects group, a technically progressive contemporary group, and kids who were ready to try anything. The combination meant that dailies would go all the way from shots that looked great to "What the heck's that?!" There was a wide range of feeling and energy in the shots. In some ways, the visual effects in *Star Wars* were a tapestry, which was pretty fresh—and the audience hadn't seen anything quite like it before.

In those days of the late 1970s and early 1980s—the period in which the first three *Star Wars* were made—preplanning was critical. Techniques had to be decided well in advance so that proper coordination and preparation could be made for the shoot. Because optical compositing was an imperfect art, there were many restrictions imposed on George Lucas and ILM. So when we started on *Star Wars*, we didn't think much about the landscapes; it was really just starships and outer space. But as we did the next two films, a growing percentage of the movie became places that couldn't be done on stage sets, locales that were completely fictionalized. With the prequel trilogy, this aspect just got bigger and bigger, with fewer and fewer real sets—just bits of rooms and chairs—completed with massive environments visualized by art departments and built as computer graphics and models.

Today, in 2005, digital techniques have, for the most part, removed the old constraints. There is far more flexibility in shooting and in postproduction to make the visual effects shots appear to be in exactly the same style as surrounding live-action shots. Visual effects plates can usually be photographed like any other part of the film. The days when the audience knew a visual effect was coming—because of the locked-off shot, or the grain buildup, or the suddenly different cinematography—are gone.

There is a potential danger, however, with these new tools. If not properly handled, digital techniques can inject an artificial presence into a movie. Today, a lot of contemporary visual effects look computer-generated because many of the artists haven't studied nature enough. The earlier groups were always trying to enhance reality, not do away with it, by creating alternate realities and different views of reality. So I've begun asking the younger artists to go outside and look at the buildings where they're working, at the cars they drove to work in, to notice how the light bounces off the cars to the ground, and so on.

John Knoll is from a generation who understands the importance of models and studying nature. He knows what it's like to build and touch a model, but many of the younger generation don't know what it's like to feel those things. If I look at something that's pure CG next to a rendered image that is real, I can see the difference in a flash, and John can see it in a flash, and the audience can see it, too—but some CG artists actually can't see it. And this is becoming prevalent in the industry.

A *Star Wars* movie is supposed to be real, with real scenes with real human beings—and creating a live-action movie that seamlessly interacts with spectacular visual effects is infinitely more complicated than doing an all-CG feature. Digital techniques have allowed the artists involved to focus less on mechanical assembly issues—is there a matte line, does the move match, is the color right?—and more on aesthetic issues—does the shot seem scary enough, is there enough energy, how well are we establishing the mood? And this is a good thing. It allows artists to spend most of their time making art, and not wrestling with technology, which translates directly into us being able to do better work with the same resources, and the audience getting a more realistic and entertaining experience. But that can't happen without an intimate knowledge of nature, because it will always remain the basis for the most fantastic imaginings.

*Dennis Muren, ASC*
*Senior Visual Effects Supervisor*
*Industrial Light & Magic*

# Introduction

The object of this book is to reveal how the environments of *Star Wars* were created, and to explain why some were done one way and some another. Over the course of the six films—Episodes IV, V, VI, I, II, and III—the techniques used to create these environments evolved quite a bit. Between 1977, when the first *Star Wars* film came out, and 2005, when the final *Star Wars* movie was released, the technical evolutions have engendered more and more freedom in the design of the shots, which has meant more ambitious environments, and the elimination of the "locked-off shot"—the curse that limited many of the earlier visual effects. The saga has gone from the first incredible innovations of motion-control photography in *A New Hope* to the refining of digital cinema in *Revenge of the Sith*, and this book tells a part of that story.

When I was kid, I was fascinated by movies that showed places, events, and objects that couldn't be real. Epic battles at sea, cities destroyed by aliens, miniature submarines navigating the bloodstream of a human body—these all caught my attention because I knew they couldn't have been shot on a location. Somebody had to have made them, as they were clearly artificial realities. But if film crews didn't photograph these things, who did? And what techniques were used to create the fantastic worlds?

Sometimes it was pretty obvious how it was done. Models looked like models. Matte paintings were easy to spot. Other times I could tell that something funny was going on because certain scenes didn't stylistically match the rest of the movie. When visual effects did blend in, though, they were magic. When they were skillfully executed, quasi-magical techniques enabled filmmakers to create illusions that could actually convince an audience that a spacecraft was landing on the moon.

Perhaps the same desire that motivates a child to seek out the tricks behind a magician's performance prompted me to search out how visual effects were done. This fascination at first led to hobbyist experimentation. I built models and photographed them to see if I could create the same cinematic illusions. It was great fun, and this hobby became progressively more serious over the years. I built more ambitious models, made clay animated stop-motion films, and generally made a mess out of my parents' basement.

I was fourteen years old when *Star Wars* came out. It had a huge impact on me because I saw it at a time when I was just starting to think about what I might want to do with the rest of my life. I was wondering which career path I should take—when suddenly a new era of visual effects began. In fact, by the age of fourteen I had convinced myself that I understood all the key visual effects techniques—but *Star Wars* was different. I remember leaving the theater not knowing how some of it had been done. The filmmakers who had made *Star Wars* were doing things in a new way, and I

wanted to know how. Because *Star Wars* revitalized the entire visual effects industry—inspiring all kinds of new, exciting projects—visual effects became a viable career path. By the time I'd finished high school, I was determined to be a visual effects supervisor, with the idea that someday I would help create something as impressive as *Star Wars*.

I also had another *Star Wars*-related life-changing experience, in May 1978. My dad was attending a nuclear medicine conference in Anaheim, so he brought my brothers and I out to Southern California. At this point Industrial Light & Magic was still down near Los Angeles in Van Nuys. I'd been reading all these magazines about how *Star Wars* was done, and there had been a great article in *American Cinematographer* with interviews with John Dykstra and all those guys about how they'd done it. On a lark I picked up the phone book and was able to find a listing for ILM. I called up and got a hold of Grant McCune who was head of the model shop, and I explained that I was a model-maker interested in doing it for a living. He said, "Why don't you come on down and we'll show you around?" So my dad dropped me off and I hung out at ILM for a whole day—and, boy, that was an awesome experience. I saw in the optical department how the elements came together; I saw models under construction in the model shop. It was all really cool and the visit made it real to me. When you see people coming in to work and that's their job, you say to yourself, I could be one of these guys!

I went to USC film school, where I think I was the only one in the film program who hadn't decided they were going to be a writer-director. I was more interested in the technical aspects of filmmaking. In the end, I'm a craftsman at heart. I didn't want to graduate and then start at entry level, so during summer breaks I worked as a model-maker. My last semester, I took an advanced animation class and built a motion control system that I bolted onto an Oxberry animation stand and made a two-minute "slitscan" movie. I learned a lot about the nuts and bolts of technical film-making and animation. After graduating, I started working as a motion control camera assistant about half the time, and as a model-maker the other half. I'd been out of school for about six months when a former instructor called and said ILM was looking for a motion control camera assistant.

My first days at ILM were an eye-opener for me. I had somehow imagined that the premiere visual effects company in the world would have all of the best equipment. My last freelance job in L.A. had been at Cinema Research, where I worked with their brand new motion-control camera. Now, here I was at ILM, working with the original Dykstraflex camera, and I was shocked to discover that the boom had about two inches of mechanical slop in it. I couldn't believe it. A small shop in L.A. had better equipment than ILM? They still did the best work in the industry. How was that possible? It became apparent to me that it was all about the eye of the artist. A good artist needs good tools, but the talent is the most important thing.

After a few months assisting on the motion-control stage, a position opened up for an operator in the Effects Camera department, where I got an opportunity to work directly with the supervisors to create elements for shots and use my imagination to solve problems. A lot of really fun challenges got dropped on me during that time. I had become the in-house slitscan expert, and when we were doing the pilot episode of *Star Trek: The Next Generation* (1987), I was tasked with figuring out how to make the *Enterprise* "stretch and snap" into warp drive. The technique I came up with worked reasonably well, and I got a kick out seeing it on television every week for years to come.

Projects like that caused me to move up in the company. Dennis Muren, Ken Ralston, and Dave Carson provided me with fantastic opportunities to learn and grow and take risks. I learned a lot of valuable skills from them, and my approach to supervising now is modeled after aspects of each of these guys that I liked the most.

My first film as a visual effects supervisor, *Hudson Hawk* (1991), was not the best experience. If your first film isn't pretty impressive you might not get another chance—and in fact I didn't get another feature until a good friend of mine, Patrick Johnson, hired me about four years later. He was directing *Baby's Day Out* (1994), brought the work in to ILM, and asked me to supervise it.

Fortunately between the two features, I'd done a dozen or so TV commercials, where you get lots of client interaction experience, which is a key element to becoming a supervisor. That, plus *Baby's Day Out*, got me back into the rank of supervisors again. A few years later, I met with ILM president Jim Morris about where my career was going, and said, "I like science fiction; if something like that comes in, please keep me in mind." A few days after, *Star Trek: Generations* (1994) came in. That went pretty well, and one job flowed into another until 1995 when George Lucas decided to tweak *Star Wars* and create the *Special Edition* series.

My experience on that project led to me being hired as one of the visual effects supervisors for *Star Wars*: Episode I *The Phantom Menace*. Now twenty years after seeing the first *Star Wars*, I was going to have a go at the *Star Wars* galaxy. I was going make many of the decisions about set building—which environments would be built as complete practical sets; which would be partial sets with visual effects to extend them; and which worlds would be entirely created by ILM. The big question is always: How much do you build? Some environments are too big and therefore cost-prohibitive to create in their entirety—but visual effects can be expensive, too. If you build only part of a set, it may be possible to play much of the scene against that one part, which makes visual effects unnecessary. But is it cheaper, and better for the film, to extend the set in postproduction? What if we don't build anything and spend the money in postproduction to create something grander than we could have with a practical set?

There are no hard and fast rules for making these decisions. It depends on what the scene is, how long it is, to what extent the actors will interact with objects on the set, and what the set is meant to be in the first place. In the pages that follow, readers will learn how many of these decisions were made by myself, the pioneers of the first trilogy, and my peers.

No matter how much we build or don't build, our goal at ILM is always the same: to do the visual effects correctly, so the audience won't know how the environments were created. They won't know which was a set, a miniature, a digital or practical matte painting, or if the set was entirely computer generated. They will be transported to other worlds and believe all the more fully in the fantastic story being told on the screen.

*John Knoll*
*Visual Effects Supervisor*
*Industrial Light & Magic*

STAR WARS: EPISODE IV

# A FILMMAKING REVOLUTION

# Founding ILM

"Special effects" usually refers to physical tricks and pyrotechnics performed on set or with miniatures. "Visual effects" generally refers to optical tricks or digital enhancements—those planned in preproduction and created in postproduction. The special and visual effects of *Star Wars*: Episode IV *A New Hope* came out of the technology of the mid 1970s. Prior to George Lucas and company's groundbreaking efforts, the state of the art of these disciplines was represented by such films as *Tora! Tora! Tora!* (1970), *Bedknobs and Broomsticks* (1971), *Silent Running* (1972), *Earthquake* (1974), *The Hindenburg* (1975), and *Logan's Run* (1976).

Lucas knew that the special- and visual-effects sequences he had devised—notably the attack on the Death Star—were so unlike anything that had been seen before that he would have to develop revolutionary new techniques to pull them off on screen. In June 1976, Lucas asked John Dykstra—who had collaborated with *2001: A Space Odyssey* pioneer Douglas Trumbull on *The Andromeda Strain* (1971) and *Silent Running*—to supervise the special photographic effects. With the backing of Twentieth Century Fox, which agreed to designate $2 million for the project, Lucas established his own special-effects house in a 30,000-square-foot warehouse in Van Nuys, California, sixteen miles northwest of Los Angeles. Dykstra was put in charge of this "miniature and optical-effects unit," which would soon come to be better known as Industrial Light & Magic (ILM).

The *Star Wars* revolution was about motion control. Dykstra had been talking about doing a project where he would perfect computer camera movement. He knew that if he could record certain moves, one axis at a time, and play them back exactly, it would be possible to do more complex camerawork than ever before. When George hired him for *Star Wars*, they both realized that this film would need such an innovative system—ultimately the Dykstraflex motion-control camera was born (below right, Dykstra watches a playback in the Dykstraflex; the model in front of him is part of the Death Star hangar wall). On a boom rig, for example, the camera could record shots of a *Millennium Falcon* miniature (opposite page, with Doug Smith leaning in, Steve Gawley on knees, Paul Huston and Richard Edlund in T-shirts). Then, thanks to a rat's nest of cables leading back to the electronics for the system (top right, Al Miller), consisting in part of handmade Stepper Motor Drivers, the camera could repeat the exact same move that would eventually evolve into never-before-seen kinds of imagery.

# Staffing ILM

George Lucas and John Dykstra assembled a team of young experts to help realize the ambitious sequences in *Star Wars*. One of ILM's first employees was designer and storyboard illustrator Joe Johnston (top right, painting the Rebel blockade runner miniature). Another original member of ILM was Dennis Muren, who was recruited as a cameraman (opposite page, in armchair center; on his right is Ken Ralston, on his left is Lorne Peterson; at the rear of the room, head on hand, is George Lucas, and to his right Gary Kurtz; behind George is Rose Duignan). Other new hires included Richard Edlund (middle right, adjusting the model-rig of a TIE fighter), Steve Gawley, and Grant McCune (below right, just beginning work on the Star Destroyer miniature).

Despite the pressures, the atmosphere at ILM was loose and casual. There was no time clock, and daily schedules were not enforced. The crew frequently worked well into the night and sometimes slept on the premises. During the summer months, it became very hot. The bright lights needed for shooting the miniatures drove the temperature inside the non-air-conditioned buildings to 120°, so the crew would sometimes take a break from the heat by cooling off in an improvised tub in the parking lot (below, Dave Berry). Ultimately, as deadlines loomed, stricter scheduling was utilized, and though ILM personnel worked long hours to finish the film on time, this was not always apparent to outsiders. Important visitors, like the occasional executive from Twentieth Century Fox, would arrive to find a number of crew members goofing off in the parking lot, and come to the wrong conclusion about their work ethic.

# "The Star Wars"

Lucas and ILM faced a formidable visual effects task. Other aspects, particularly environment creation, proceeded in a more conventional fashion, combining the physical sets built at Elstree Studios in England with the relatively standard matte painting and miniature techniques of the time.

As Lucas went through various drafts of what was then being called *The Star Wars* (from 1972 to 1976), preproduction continued (opposite page: a production blueprint for the *Millennium Falcon* cockpit set). However, in the back-and-forth budget negotiations with Twentieth Century Fox, some locations were combined to cut costs, presenting new challenges to John Barry, the production designer George had hired. Consequently, all sequences on the planet Alderaan were eliminated from the film, with the prison scenes that took place there relocated to the Death Star. Given the costcutting, many of the props were actually made out of junk. And in fact a lot of the work done on *Star Wars* was very seat-of-your-pants (right: in a makeshift room, John Dykstra and Grant McCune are at work in "editorial," with a custom-built VistaVision Moviola on the left and a 35-millimeter Moviola at rear; a trim bin is next to the sawhorse table).

***Star Wars: A New Hope*** (1977)

WRITER-DIRECTOR: George Lucas
PRODUCER: Gary Kurtz
PRODUCTION DESIGNER: John Barry
DIRECTOR OF PHOTOGRAPHY: Gilbert Taylor
PRODUCTION SUPERVISOR: Robert Watts
PRODUCTION ILLUSTRATOR: Ralph McQuarrie
ART DIRECTORS: Norman Reynolds, Leslie Dilley
SET DECORATOR: Roger Christian
EDITORS: Paul Hirsch, Marcia Lucas, Richard Chew
SPECIAL PHOTOGRAPHIC EFFECTS SUPERVISOR: John Dykstra

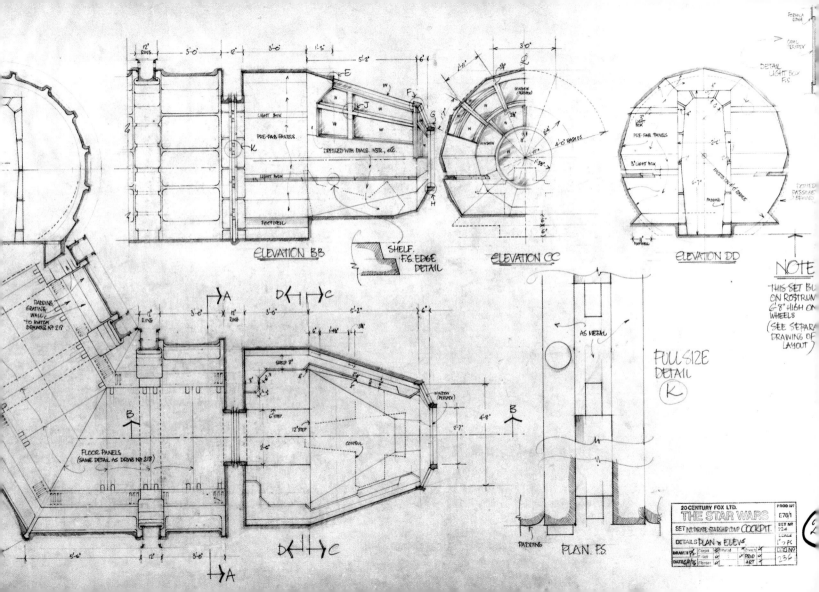

# Rebel Blockade Runner 1

The opening shot of *Star Wars* began the saga with just the right impact. The idea was brilliant: We see a spaceship pass overhead. It's being fired upon, and a moment later we see by what: a larger ship, but it's even larger than we think. Much larger. It just keeps coming. *Wow, this thing is HUGE!* When the ship finally finishes passing overhead, we know the little ship is doomed. This impressive shot amazed audiences all over the world, and was even more impressive  because it was done with a three-foot-long model.

The smaller spaceship was called the Rebel blockade runner, aka the *Tantive IV*. These blockade scenes—in which the droids converse, Rebels exchange fire with stormtroopers, and Darth Vader (David Prowse, right) enters—are a good example of a technique used frequently in filmmaking: multiplying one environment into many, by repeatedly using the same set. In reality, only two sets were built for all of the opening sequences, but clever photography and editing makes them look numerous.

One of the two constructed sets was L-shaped, and it was used for all of the ship's corridors. This set is seen dressed two different ways: once with a door at the end of the long leg—which the stormtroopers come blasting through (below)—and once with the door replaced by a right turn, implying another corridor (opposite page, top). The set of the escape-pod bay is also used several times as multiple areas (opposite page, below).

Rebels take up positions facing the door. Later they flee stormtroopers down the same set, which has had its door replaced with a right-hand turn.

R2-D2 and C-3PO in the corridor set. Lucas makes good use of the turn at the end. Rebels run around the corner and, a few shots later, are seen running around the same corner in a mirror image of the exact shot. In fact, every character in the scene rounds this corner at least once.

Captured Rebels are marched down the modified corridor. Later, Darth Vader confronts Princess Leia in the same corridor, then walks past camera, turns the corner, and walks down the corridor again as he speaks with his aide.

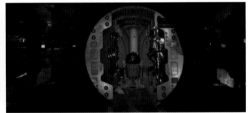

C-3PO in the escape-pod bay. Leia entrusts R2-D2 with the plans for the Death Star with the camera looking down one direction of the bay. The droids then head off in the other direction.

Darth Vader interrogates a Rebel at the far end of the bay. Stormtroopers search for Leia on the same set. A few feet farther, R2-D2 and C-3PO make their escape on the same escape-pod bay set.

## Rebel Blockade Runner 2

Principal photography for *Star Wars* began in Tunisia on Monday, March 22, 1976. Location shooting wrapped on April 4, and studio shooting commenced on April 7—continuing up to July 16, winding up an eighty-four-day shoot. Lucas shot on the blockade-runner set during those last few days in July. The escape-pod bay set, in particular, had to be built in a hurry (opposite page). But because it was last, good use was made of salvaged bits from other sets that had already been destroyed after the filming of their scenes was completed. The engine-like large objects that flank the escape-pod bay's walkway came from the set of the Death Star chasm (below left, hanging from ceiling). Likewise its floor gratings came from the prison cell-block set (below right).

# Tatooine Desert

The Tatooine dune sea scenes were shot during the first few days of location shooting in Tunisia, which lasted from March 22 to April 4, 1976. Many of the props and costumes were being used for the first time. After their escape pod lands on the desert planet (opposite page, with two crew members), the two droids—C-3PO (Anthony Daniels) and R2-D2 (Kenny Baker)—wander around the desert (right) before they're both abducted by Jawas. The C-3PO suit had a painful tendency to pinch Daniels as he walked, and the radio-controlled R2-D2 refused to roll. The solution was to have a crew member put the droid on skis and pull it along by an off-camera rope. Stormtroopers (being directed by Lucas, below) are on the droids' trail.

# Sandcrawler 1

The sandcrawler, an enormous tank-treaded vehicle used by the tiny junk-dealing Jawas for transportation and shelter, was made in part by building a full-scale section of the bottom portion of only one of its sides. This was used for close-ups, and was planned for use at two locations: at the Lars homestead (top right, with Luke's speeder on a welded steel post), where Luke Skywalker (Mark Hamill) and Uncle Owen (Phil Brown) purchase the droids; and, representing two film locales, in a canyon—for a shot of the Jawas carrying R2-D2 to the sandcrawler (opposite page), and for the scene in which Luke and Obi-Wan Kenobi (Alec Guinness) find the slaughtered Jawas (below right, with dead Jawas and Luke's speeder on a cantilevered rig; the sandcrawler is redressed with blaster scorch marks). The set was constructed so it could be disassembled and hauled on a series of trucks to each location. However, after the homestead shoot wrapped on Thursday, March 25, 1976, the sandcrawler was one of the casualties of a freak desert storm. The set was literally blown apart, with pieces scattered hundreds of yards away. The shooting schedule was therefore shifted around to allow the art department time to collect the pieces and rebuild the vehicle for the canyon scenes on Tuesday and Wednesday, March 30 and 31, 1976.

## Sandcrawler 2

A 1.6-yard-long radio-controlled model of the entire Jawa sandcrawler was used whenever a wide shot was needed. For the scene in which the crawler transitions from mountainous terrain to desert sand dunes, the miniature (top right, Paul Huston and Joe Johnston painting the model) was taken to Death Valley, California, where the terrain was a good match for that of Tunisia (opposite page). Being careful not to spoil the sand and mud around the crawler model with out-of-scale footprints, Bill Shourt and Grant McCune at least once had to use a fully extended light stand to give the sandcrawler a gentle nudge in the right direction (below right). Note that the two—practical set and miniature—were almost never combined in a single shot. The one exception is of the Jawas carrying R2-D2 to the crawler (below left, final frame). Sleight-of-hand editing eliminated the need to show them around the miniature sandcrawler.

## Stormtroopers BTS

A rare panoramic photo taken behind the scenes during the location shoot in Tunisia.
A stormtrooper is touched up by a crew member, while Lucas (next to camera) studies
the framing (below, C-3PO on Tatooine).

■ 00405128

# Lars Homestead 1

Lucas filmed the Lars homestead exteriors in two different Tunisian locations. The above-ground portion was shot on a dry lake bed near Nefta (opposite page, with base camp, sandcrawler, and homestead "igloo;" below right, with homestead burning); the sunken-pit part of the home was filmed at the Hotel Sidi Driss in Matmata (below, third from left). Editing makes these two disparate locations appear adjacent to each other, when Luke Skywalker runs over to say something to his Aunt Beru (below). The lip of the crater was merely a raised berm of earth (top right, with sandcrawler); there was no pit at the dry-lake location.

Not all of the homestead exterior shots were filmed in Tunisia, however. A matte painting of the homestead and its surroundings was created at ILM for Luke's electrobinocular point-of-view shot (middle right, with Tatooine background on cylindrical surface). The decision to insert the POV occurred after shooting was completed, and it was easier to create this shot with a painting rather than go back to Tunisia.

## Lars Homestead 2

The area surrounding the Hotel Sidi Driss in Matmata was not very Tatooinesque (note the palm trees, opposite page), but the sunken main courtyard made a unique home for Luke Skywalker. The courtyard was the largest of the underground hotel's five open areas, with a bar and a restaurant (below left; just off camera is a stage used by performers and musicians entertaining clients of the restaurant)—and was much larger than the lake-bed crater circumference (below right). The restaurant was made up of little alcoves. The dining area where Luke and his guardians, Uncle Owen and Aunt Beru (Shelagh Fraser), share a meal and discuss his future was actually one of the real alcoves (top right; to the right are a few steps that lead to a closet in the real world, but to the Lars homestead kitchen in the film).

## Lars Homestead Interiors

Two homestead interior sets—the garage and the kitchen (the dining nook is part of the
Hotel Sidi Driss)—were built in London, on Stages 7 and 1 of EMI Studios (Elstree) at
Borehamwood. The scenes shot on these two sets constituted the first days of studio
photography on April 7–9, 1976. The garage, where Luke cleans up the droids and stumbles
upon Princess Leia's hologram message, is a partial set (below). Only about 270 degrees of it
was built; the rest was left open for camera access. The two doors in the back wall of the
garage show a parked landspeeder and a skyhopper (opposite page). The kitchen was a complete set; in the film, it's supposed to be linked to the dining area by a small stairway (right).

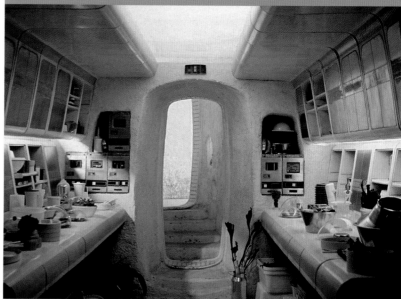

# Luke's Speeder

Luke Skywalker's landspeeder is supposed to hover a few inches above the ground as he and C-3PO go off in search of the zealous droid R2-D2. When the landspeeder was under construction (middle right), it originally had a full canopy.

The levitation effect was achieved three different ways: First, the speeder was balanced on a long counterweighted arm that could pivot around its center (below left, with Lucas on right). This rig was used at the Lars homestead and at the wrecked sandcrawler site for shots of Luke getting in and out of the speeder, as well as whenever he pulled away. Second, the speeder was supported by a welded pipe (below right). This was usually framed out of shot, or hidden behind an object like R2-D2 or a rock. Third, a wheeled speeder was used for whenever Luke was seen driving (below middle). When the shots were wide enough that the wheels would show, a mirror skirt was added (opposite page, in postproduction, with body doubles for Luke and Obi-Wan). This effect was fairly apparent when seen up close, but convincing in wide shots (top right, final frame).

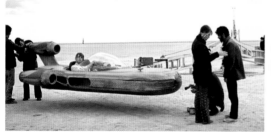

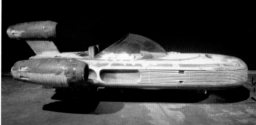

## Ben Kenobi's House

R2-D2 leads Luke Skywalker to an old hermit, Ben, who turns out to be the legendary Jedi Knight Obi-Wan Kenobi. His house was another complete, albeit small, set that was built on Stage 7 and shot on April 23 and 24, 1976 at Elstree (top right: Lucas, lightsaber in hand, discusses the scene with Guinness, as Daniels and Hamill look on).

Kenobi has an interesting collection of objects—one of them looks like a stuffed armadillo (opposite page, left). The trunk (opposite page, right) contains the lightsaber of Luke's infamous father, Anakin (aka Darth Vader), and the rug is suspiciously bearlike. Kenobi eventually offers the lightsaber to Luke (below, without lightsaber visual effect).

## Mos Eisley 1

After Luke Skywalker makes the decision to accompany Ben Kenobi to the planet Alderaan, they travel to a "wretched hive of scum and villainy." Mos Eisley street locations were all filmed in one small area on the island of Djerba just off the coast of Tunisia. This block of buildings was dressed with vaporators, antennas, and spacecraft wreckage. In fact, flat, painted cutouts of Tatooine-style abodes were placed in the far background, while one particular crashed starship covers up a modern building that didn't fit with the "Tatooine" style (opposite page, center right). Upon arriving, Luke and Ben are stopped and questioned by stormtroopers (below and right). Because the droids had a tough time rolling on sand, whenever their wheels were off camera—as in this particular shot—the crew would lay down plywood panels for easier maneuverability (right: note the modern buildings at the very top of frame).

## Mos Eisley 2

A close-up of the big, crashed ship that covers up the modern building also reveals a tree behind it (below right). In another Mos Eisley street, just outside the Cantina next to the dewback, is a furry, dressed creature that was perhaps originally a llama (top right). Another spaceport oddity is a parked "flying saucer" (opposite page), with gigantic landing gear attached to spindly legs (below left: another strange vehicle).

## Mos Eisley Cantina

The Mos Eisley Cantina (top right, as a maquette) scenes were originally shot over a four-day period in London (opposite page). Stuart Freeborn, the makeup and creature supervisor, was ill during the prep for that shoot, and his absence compromised the richness and variety of the alien masks. Back in Los Angeles after principal photography had wrapped, Lucas had additional aliens created. With his friend and filmmaker Carroll Ballard, he shot many of the most memorable characters on a tiny soundstage in Hollywood with just a small fragment of a set (below right, with Ballard behind the camera and Lucas standing). As part of the same shoot, all of the Greedo close-ups were reshot with a more elaborately articulated face mechanism (below left: note the high-heels—the bounty hunter was played by a woman). Careful matching and editing make all of the cantina footage appear to be from the same shoot.

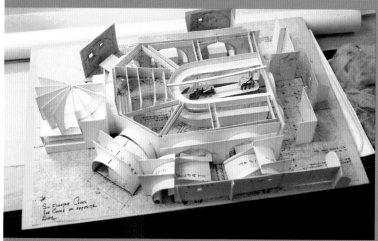

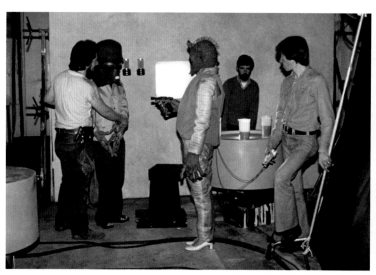

## Mos Eisley Alleys

The back alleys of Mos Eisley were built on Elstree stages because no suitable Tunisian location was found during scouting. Ultimately, the studio interior was a much more hospitable and controllable environment than that of Tunisia. By keeping the backstreets narrow and enclosed, the lack of vistas wasn't a problem (below and opposite page: two views of the same set).

The alleys are where the droids hide from stormtroopers, where Luke Skywalker discusses the sale of his landspeeder, and where the heroes rush off to Docking Bay 94, having met Han Solo (Harrison Ford) and Chewbacca (Peter Mayhew). At the entrance to the docking bay, the number "94" is visible (opposite page, upper right; also note the two R2 units).

## Docking Bay 94

Like the *Millennium Falcon*, Docking Bay 94 was only half built. The large circular set was too large to fit on a soundstage but, because the bulk of the *Falcon* could be counted on to obscure the docking bay's missing side, George decided that half a set would be enough (opposite page, the set and *Falcon* under construction). Two of the scenes shot here were Han Solo's negotiations with a human Jabba the Hutt (right: this scene didn't make the final cut, but was later re-created with a digital alien Jabba for the *Special Edition* of *A New Hope*) and Han's shootout with stormtroopers (below).

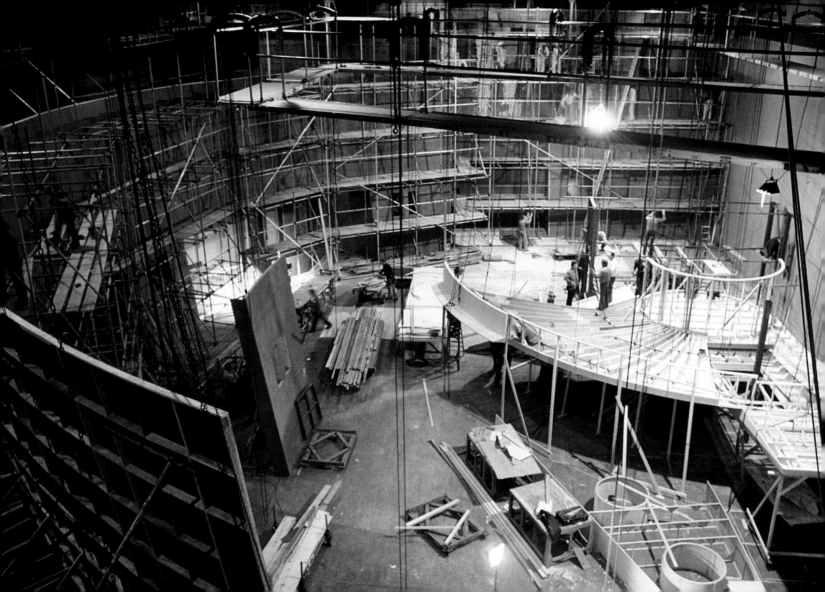

## Millennium Falcon Exterior 1

Because the *Millennium Falcon* would be seen from only one side throughout the film, only one portion of the starship was constructed (below right). The *Falcon* was nevertheless too large to fit through the soundstage door. Therefore, after shooting was completed on Docking Bay 94 (below left), the set was removed and the Death Star hangar was built around the remaining *Millennium Falcon* set (opposite). In both cases a little false perspective was also used: The half-*Falcon* is much closer to the wall than it could be if it were whole. This trick isn't visible in most cases—except when the camera was higher. For one such instance in the Death Star hangar, a matte painting (top right) completed the ship and made the area appear larger.

# Millennium Falcon Exterior 2

A panorama of the *Falcon* exterior, with the actress playing Greedo standing next to the ship (opposite; below, boarding the *Falcon*).

## Millennium Falcon Interior 1

The *Millennium Falcon* interior set (below right, under construction, seen from the exterior) consisted of an entrance ramp, a curved corridor with a ladder leading to the sole gunport, a branching corridor leading to the cockpit (built on another stage), "smuggling" compartments (below middle, final frame), and a main room (opposite page) with a chess table where Luke Skywalker trains with a lightsaber against the remote (below left, Joe Johnston operates the remote at ILM; middle bottom, final frame).

Before construction began, a foam-core model of the interior set was created (top right). Models like this one are almost always built by the art department to confirm for the film's director that the set will meet all the requirements. On the final set, the wall opposite the round corridor of the main room (opposite page, right) was left incomplete for camera access. When this set was rebuilt for *The Empire Strikes Back* (no one had anticipated the success of the first *Star Wars* film, so the original set had been destroyed), the missing wall was finished and the walls all became "wild"—that is, removable for camera access.

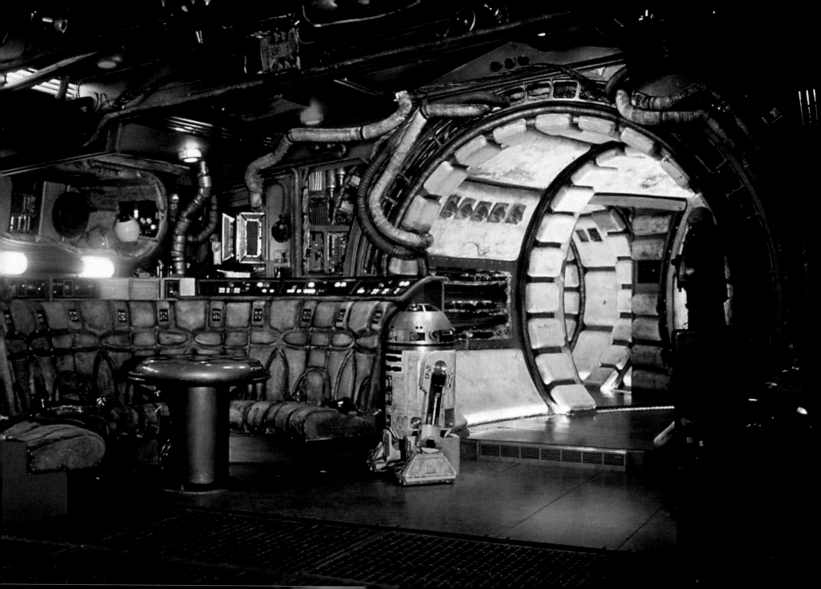

## 24

## Millennium Falcon Interior 2

The interior of Han Solo's *Falcon* was never filmed from the outside. Whenever an actor ran up the exterior access ramp, he encountered a dead end (top right); his continuing path would be picked up on the interior set's ramp (below right), and the two actions would be edited together. Other clever editing covered up the fact that there was only one gun port, which was used for both Han Solo and Luke Skywalker as they fight off TIE fighters (opposite page and below left).

## *Millennium Falcon* Cockpit

The *Millennium Falcon* cockpit was not built adjacent to the *Falcon*'s main interior set for a handful of reasons (opposite page, the unattached cockpit): First, the script didn't require them to be next to each other; it was sufficient to see an actor turn down the set of the *Falcon* hallway and, following a cut, walk onto the cockpit set. Second, the cockpit needed to be raised eight feet from the stage floor to allow enough clearance for the bluescreen, which was necessary for shots of the actors looking out into space (below). Third, the set was constructed on an "inner-tube rig"—that is, stacked on a number of truck tire inner tubes, so the entire cockpit could be shaken by stage hands to simulate explosions rocking the ship. The cockpit set was just small enough to be mounted on one of these rigs (right). The other interiors were too large, so side-to-side movement was created by shaking the camera and body language from the actors.

# Death Star Laser Tunnel

The Death Star's fearsome planet-destroying weapon is used only once—to obliterate Princess Leia's home planet of Alderaan. The laser tunnel began life as the "small thermal exhaust port" seen at the end of the film—the Death Star's one vulnerable point, which Luke Skywalker exploits to the Rebel Alliance's advantage (below left).

To repurpose the port as the Death Star laser tunnel, the model was redressed slightly with the addition of illuminated cylinders at the top and bottom (below right). The eight-inch tunnel was made to look miles long through the use of mirrors: one mirror capped the end of the tunnel, and the camera shot through a second, half-silvered mirror. The two parallel mirrors create the "infinite tunnel" effect (top right, final frame).

When the weapon is fired, a laser beam shoots down a tunnel, and two technicians filmed in postproduction at ILM react to its brightness (opposite page, with Jonathan Erland and Joe Johnston as the extras).

# Death Star Hangar

The Death Star hangar—into which the *Millennium Falcon* is pulled by a tractor beam—was built as a partial miniature (top and middle right) and as a large set (opposite page). The miniature was used for shots of the *Falcon* entering and exiting the bay. Wide shots (below left and right) used matte paintings to complete the missing parts of the set, to make the bay appear larger, and to complete the illusion of the deep freight elevator.

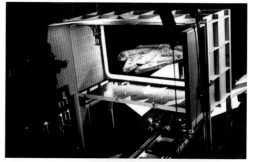

## Death Star Elevators

Many Death Star interior sets were required for *Star Wars*, the majority for only a handful of shots. Because it was natural to expect that these interior spaces would all have a similar feel, Lucas decided to make the sets modular. The walls were fiberglass panels that could be attached to scaffolding and rapidly assembled into whatever configuration was needed. Consequently, one stage with a shiny black floor became the elevators, a chasm, the tractor beam control center, and the gun bay. The elevator set (right) was made to appear much larger by extending the curved wall down below the floor with a matte painting (below; opposite page: concept illustrator Ralph McQuarrie plots the perspective of that painting extension).

## Death Star Cell Block

The cell block is another set that benefited from modularity (top right, set blueprint). The elevator that Han Solo, Chewbacca, and Luke Skywalker exit into the cell block (opposite page, behind them, as they shoot it out with the stormtroopers) is the same set from the previous scene, which saved production the expense of building it twice. The elevators themselves were not moved; the walls around the first elevator set were struck and the cell block (below, final frames) was built around them.

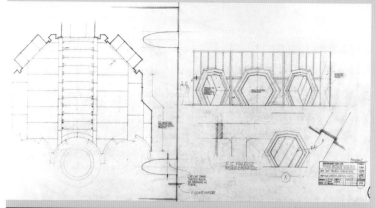

## Death Star Garbage Masher

Escaping from the shootout, Princess Leia (Carrie Fisher) leads Han, Luke, and Chewbacca into a garbage chute. The garbage masher was a three-walled set, with the two side walls rigged to slide toward each other (top right). The dianoga tentacle that grabs Luke's leg can be seen on the floor (opposite page, bottom left), in a photograph most likely taken the day after the shoot. An insert of the creature's head was shot later at ILM (below left and right, with Phil Tippett).

In Lucas's initial draft of "*The Star Wars*," Han and Luke encounter the "dai noga," a "spiderlike" monster in a hallway within the planet of Alderaan: "They are not natural creatures," Han says. "They are very large [...] Guns are no good. Only a Jedi, strong with the FORCE, can stop it."

## Death Star Chasm

The chasm set was used as two different locations: where Luke and Leia are briefly trapped but swing across to safety (right); and as the power trench where Obi-Wan Kenobi switches off the tractor beam, for which the lights on the walls were gelled green (below middle). Both scenes took place over a floor only a couple of feet below the protagonists (opposite); matte paintings were added later, creating depth, danger, and excitement (below, final frame; bottom, matte painting for the power trench).

## Death Star Hallways

With all of the chasing and running around inside the Death Star (below left, final frames), Lucas needed an inexpensive and flexible way to create the various hallways. Again, a modular approach was used. The walls were all fiberglass panels eight-feet-wide by ten-feet-high. As such, they could be rapidly rearranged to create a corner, an intersection, a corridor, or anything else that was needed (right: Darth Vader and an Imperial officer stroll purposely down a generic Death Star hallway).

*Star Wars* films often contain environments, like these hallways, with shiny floors—and they're always a bit of a headache. No matter how careful the crew was on *A New Hope*, always wearing white booties, the floors quickly revealed dirt that would be visible on film—therefore they had to be cleaned constantly (opposite).

## Rebel Base

The exterior of the Rebel base on Yavin 4 (top right, final frame), where our heroes join up with the Rebel fleet, was shot in two locations. The Massassi Temple exterior was filmed at a Royal Air Force base—Cardington Air Establishment in Bedfordshire, England (opposite page)—and completed with a matte painting (below left). The vista views were shot from the upper landing of Temple IV in Tikal, Guatemala, by Richard Edlund and Dick Alexander of ILM. Model-maker Lorne Peterson played the Rebel lookout, whose perch was erected on the site (below right, with Richard Edlund).

## Rebel Briefing Room

In the briefing room, the Rebel pilots are told the details of the Death Star attack mission (right). Behind them the transparent maps of the war room can be seen (opposite page), where the progress of the attack is monitored a few scenes later (below, final frames). It was logical to combine the sets this way, as it saved costs to build the two on the same stage.

On the briefing screen, early computer graphics are visible. I remember in 1978, while I was still in high school, a computer graphics creator came to the University of Michigan and gave a guest lecture about the burgeoning technology, including shading and rendering solid surfaces. He showed a computer-generated strawberry and, at the time, I thought, "Wow! Look at that strawberry! It looks solid and has a glossy highlight!" The graphics displayed during the Rebel briefing, like the strawberry, were just the beginning of the computer's foray into filmmaking.

# Rebel Base Hangar

The Massassi Temple hangar set made clever use of a few tricks that served to enlarge the environment. Strings of lights in front of a black curtain made the far wall appear to be much farther away than it actually was (opposite page, on right). And, because only one full-sized X-wing and Y-wing starfighter props were built for the film, all of the background X-wings and Y-wings were cutouts (below left, full-sized Y-wing is in foreground—behind the full-sized X-wing, with ladder, are the cutouts; the flat nature of the background ships is apparent; they are also stacked closer together than would have been possible if they had wings, but their inclusion visually increased the hangar's size). One wide shot of the hangar was created by taking one plate of the set (below left), and creating a matte painting (top right) with two holes in which that same plate is placed twice (below right).

# X-wing Starfighter

Three X-wing starfighters were created for the original *Star Wars*. One was a full-sized exterior built for the hangar scenes on Yavin 4 (below left). This set was used for all the shots of pilots climbing into or out of the fighters, and was therefore constructed in a wings-closed, landing-gear-down configuration.

The second starfighter was a simplified half of an X-wing, which was built for all of the cockpit scenes (below middle and opposite page). It had engines but no wings. The one cockpit set served as the interior for all the fighters. Because the second set was built with its wings in the open configuration, a minor continuity error can be seen in the film: The squadron's wings are already open before Red Leader (Drewe Henley) gives the order to "Lock S-foils in attack position."

The third X-wing set consisted only of the front half of the cockpit, for character point-of-view shots (below right).

# Y-Wing Starfighter

As with the X-wing, a Y-wing full-sized exterior was built (right), as well as a cockpit set that pulled apart for front, side, and back views (opposite page, set; below, final frames). Additional miniatures were built for explosions and "hero" shots.

## TIE Fighters

The only full-sized portion of a TIE fighter built was the cockpit (opposite page), which could be split in half, front to back, for camera access. The same piece was used both for the standard TIE fighters and for Darth Vader's specialized TIE fighter. Throughout principal photography, Vader was to fly a TIE fighter with the same design as the others (top right, standard). In postproduction, however, Lucas decided that in the shots showing three TIE fighters, it would be easier to understand which was Vader's ship if it was unique. The result is a small continuity error, for Darth Vader's TIE fighter cockpit appears to have a back window (middle right), but the new miniature shows no place for a back window (below right)—the result of a mismatch between the cockpit set and the miniature. Also, the set window spokes aren't in the same position as the miniature's—the latter has a spoke in the noon position (below middle), but on set that spoke obstructed the point of view from the cockpit, so it was moved slightly to allow for filming (below left). The large instrument panel in the miniature (below middle, seen through the cockpit window, which is the same in Vader's specialized TIE and in the standard TIE fighter) is not present in the set version (opposite page).

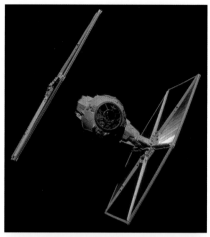

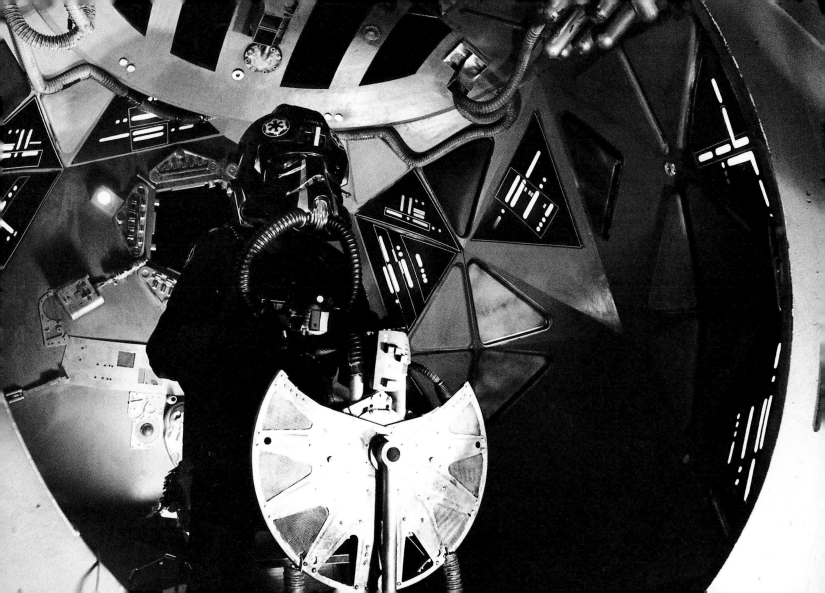

## 39

## Death Star Control Room

The control room where Grand Moff Tarkin (Peter Cushing) commands Death Star operations was a minimalist three-walled set, but was originally intended to have only two walls: the display screen and the left wall. On the morning of the shoot, however, Lucas decided that the blocking worked better if he had a right wall as well. This was swiftly built from the modular fiberglass panels, and most of the scenes in the control room ended up being photographed against this new wall. To save time and money, the consoles were reused from the prison cell block set (below left, prison cell; below right, in control room—crew record footage of a Death Star technician ultimately dropped from the film; as he's off camera, Cushing is without his boots).

Animated displays were to be projected onto a large screen that was prominently featured (opposite page), but the graphics were not ready in time for principal photography. Most of the views of the display screen are thus clean point of views (i.e., no actors) created entirely at ILM (top right). Only two shots show an actor on the set with a visible graphic, and these were done later as optical composites (middle right).

## 40

# Death Star Gun Bay

Except for the weapons themselves, the Death Star gun bay (opposite page) was assembled out of portions of other sets. The walls and the instrument tower came from the chasm set (below right), while its gun ports were the big blast doors leading to the Death Star hangar (below left).

# High-Altitude Death Star

High-altitude shots of the Death Star surface required an enormous amount of detail in the model (opposite page) and it was therefore, potentially, a very time-consuming one to build. However, a clever solution was found that allowed the surface to be built quickly and inexpensively. Still photos were taken of the low-altitude Death Star surface (see spread 42) and hundreds of prints were made. These were then cut up and tiled together (below left and right) to achieve believability (top right and middle, final frames).

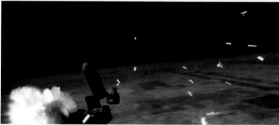

# Low-Altitude Death Star

The low-altitude views of the Death Star were created by first building masters of six different square tiles at three different scales (below are four of the tile blueprints). These masters were molded (bottom right, Pete Gerard), and thousands of copies were then cast in urethane foam and assembled as the Death Star surface on very simple plywood forms (opposite page, Death Star trench; bottom left, Luke's point of view of a fellow pilot's demise is recorded at ILM).

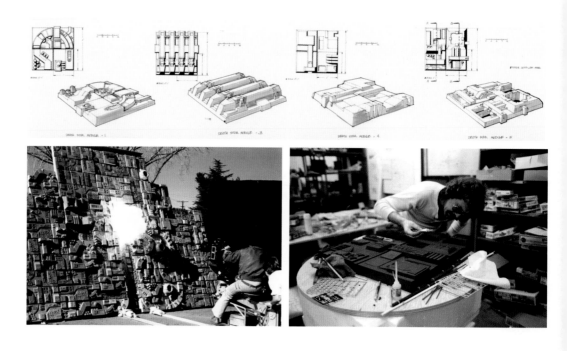

## Distant Death Star

Distant views of the Death Star were originally planned as a matte painting (top right). When initial test shots of this painting were deemed to be not realistic enough, it was decided to build a model instead (below right). A three-foot-diameter construct was created from two blown Plexiglas hemispheres. In order to produce the thousands of tiny lights on the surface, the clear Plexiglas was opaqued by painting it silver (opposite page: silver paint is much better at blocking light than black paint); after the final surface details were added and the final paint job finished, light points were added by scratching the model down to the clear plastic. A bright stage light was then positioned so it shined through all of the tiny scratches.

The Death Star originally had its laser eye along the equator, as seen in the matte painting. When this approach was abandoned and a model built, the laser eye was moved above the equator. However, the computer graphics for the briefing-room scene were already complete, and it was too late to change them, so the original design can still be seen in the film (below left, final frame).

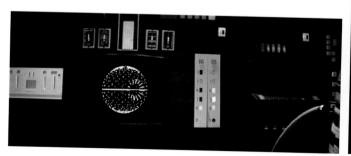

## Massassi Temple Ceremony

The set for the celebratory Massassi Temple medal ceremony (opposite page) was one of three locations built at Shepperton Studios in England (below left is a scale reference image taken at Shepperton and probably sent to EMI—where the majority of main-unit photography was being done—for Lucas's approval). All of the available stages at Elstree were already in use (the Rebel hangar and war room were the two other sets constructed at Shepperton). The set was designed so that all of the coverage could be done without visual effects, with the exception of one wide shot to show the entire room (below right) for which a matte painting was done (top right). Maximum use of the hundred-plus extras was made by moving them around and grouping the actors to fill up the background wherever the camera was looking.

## 46

## Moving North

With the success of *Star Wars* and Lucas's determination to produce the sequel in northern California, Industrial Light & Magic was moved to San Rafael in Marin County. John Dykstra, Grant McCune, Jon Erland, Bob Shepherd, and a few others decided to keep the original building in Van Nuys and start their own company there, called Apogee. Dennis Muren (below left, checking the lighting on a Star Destroyer miniature; below right, with Don Dow examining the underbelly of the Destroyer), Richard Edlund, Ken Ralston, Steve Gawley, Paul Huston, Joe Johnston, and Lorne Peterson (opposite page: Johnston in middle, Peterson on right, with masks, working on a hero model for the scenes in which the *Millennium Falcon* "parks" on a Star Destroyer), among others, made the move north to continue with ILM.

The *Empire Strikes Back* was a more complex and ambitious movie than *A New Hope*, and a great deal of additional equipment was needed. New motion-control systems, cameras, and optical printers were built to handle this new workload—the sets would also reflect the film's larger scope and aspirations. Not a big fan of directing in the first place, George decided he would oversee the film and its ongoing visual-effects revolution from an executive-producer vantage point; he also wanted to devote himself to other projects and films (such as *Raiders of the Lost Ark*, 1981) while working on Episode V, therefore he hired Irvin Kershner to direct *The Empire Strikes Back*.

*The Empire Strikes Back* (1980)

EXECUTIVE PRODUCER: George Lucas
DIRECTOR: Irvin Kershner
PRODUCER: Gary Kurtz
STORY: George Lucas
SCREENPLAY: Leigh Brackett, Lawrence Kasdan
PRODUCTION DESIGNER: Norman Reynolds
DIRECTOR OF PHOTOGRAPHY: Peter Suschitzky
EDITOR: Paul Hirsch
SPECIAL VISUAL EFFECTS: Brian Johnson, Richard Edlund
DESIGN CONSULTANT AND CONCEPTUAL ARTIST: Ralph McQuarrie
ART DIRECTORS: Leslie Dilley, Harry Lange, Alan Tomkins
SET DECORATOR: Michael Ford
EFFECTS DIRECTOR OF PHOTOGRAPHY: Dennis Muren
ART DIRECTOR VISUAL EFFECTS: Joe Johnston
MATTE PAINTING SUPERVISOR: Harrison Ellenshaw

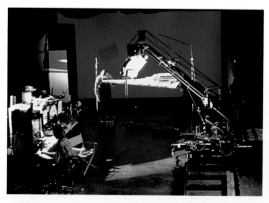

## 47

# Hoth Wastelands 1

The exteriors of the planet Hoth—where Luke Skywalker (Mark Hamill) catches sight of a probe droid (below right, with Joe Johnston) and is attacked by a wampa ice creature, and where an epic battle between the Empire and Rebel forces takes place—are a combination of live-action footage shot in Finse, Norway, photography at Elstree Studios (on stages and the backlot), and environments created in their entirety at ILM.

The choice to make whole environments rather than composite elements into live-action plates was made for both technical and aesthetic reasons. Optical bluescreen compositing is easiest against darker backgrounds, but much of the ice-and-snow-planet Hoth was going to feature relatively light objects against light backgrounds. In this situation, it would have been very difficult to avoid creating a light or dark edge around the object being composited. At ILM, on the other hand, a model was sandwiched between two matte paintings, and everything was photographed at once, requiring no compositing at all (opposite page—the miniature tauntaun with rider resides on a table hidden between a foreground matte painting on glass and a background matte painting hanging on a wall—with Dennis Muren, Phil Tippett, and matte painter Mike Pangrazio; the final frame is top right).

Aesthetically, choosing to create "whole" environments meant that the different elements could be controlled by the artists and technicians. Lighting, contrast, color, and composition became tools rather than constraints on the work. This approach, or variations on it, was used for the majority of the tauntaun and Imperial walker shots. There were very few cases when compositing walkers or tauntauns into live plates shot in Norway was required, but these instances were kept to a minimum (middle right: a wide, not-locked-off helicopter plate was shot in Norway; the plate's camera move was then painstakingly matchmoved so that the go-motion animation of the miniature tauntaun running across the snow would fit—creating what was at the time an outstanding piece of visual effects).

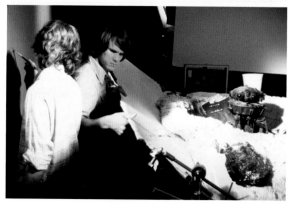

## Hoth Wastelands 2

From March 5 to 11, 1979, first-unit shooting in Finse, Norway, was slow and difficult and very, very cold. Director Irvin Kershner, actors, and crew suffered weather conditions that were harsh and uncooperative (opposite page: filming Luke on a tauntaun—note the wampa arm prop on far left; below right: shooting Luke being rescued by Han Solo—played again by Harrison Ford). Many shooting days were lost because of dense snowstorms. The progress reports list sickness, frostbite, and accidents—broken ribs and leg fractures, the latter when a skidoo (snow mobile) went over a ridge due to a blinding snowstorm (top right: this convoy proceeds in better visibility). Production fell behind schedule and some of the less scenic shooting was moved to Elstree Studios in London (below: Luke after being mauled by the wampa snow monster; amusingly, a crew member is walking past camera with the wampa's feet), where weather conditions were more predictable. The second unit continued filming background plates and inserts in Norway through April 3.

## Wampa Cave

The first day of studio shooting was on March 13, 1979. One hundred and forty-four work days later, principal photography wrapped on September 24, 1979—fifty-five days over budget. As filming took place over the spring and summer, it was not cold in England's Elstree Studios, where the cave of the wampa snow monster was built (opposite page and below right, with camera crew), and no measures were taken to cool the stages. Scenes like this were shot early in the schedule, so the stages were still reasonably cool, but you won't see any of the actor's breath in these scenes (below left, final frame)—as opposed to those shot in the freezing weather of Norway.

## Echo Base Medical Center

The Echo Base medical center is where Luke Skywalker receives treatment for his wampa-inflicted injuries and exposure to Hoth's harsh cold. The center, built on Stage 1, contained a treatment room and a recovery area (opposite page, with Carrie Fisher returning as Princess Leia; Anthony Daniels, as C-3PO, is wearing pants because only his upper half is in the shot). The former featured a three-foot-diameter "Bacta tank," where Luke is suspended in a healing fluid. A wet-suited stuntman was used to check the water to confirm everything was in working order (below left) before actor Mark Hamill was submerged (below right). The wampa scene was added to *The Empire Strikes Back* after Hamill underwent surgery following a car crash during postproduction of *A New Hope*.

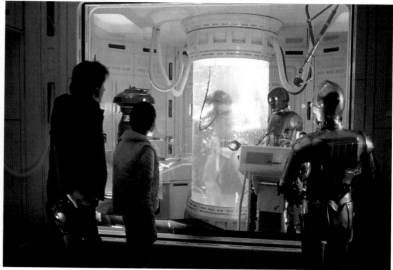

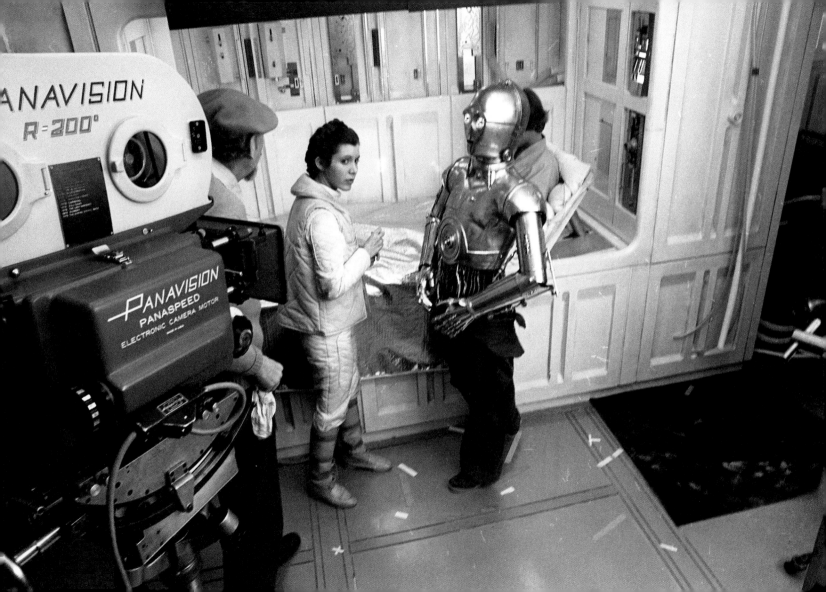

## Echo Base Hangar 1

The Echo Base hangar was a very large set—so large, in fact, that a special stage was built for it at Elstree Studios in Enlgand. Known as the "*Star Wars* Stage," it contained only a fraction of the hangar seen in the movie (opposite page; this stage was eventually torn down along with several others to make way for a shopping mall; today the reduced Elstree nevertheless has a "George Lucas" building). The ceiling (bottom right) and a much larger second hangar that housed the transport ships (bottom left) were created with matte paintings (below left: Joe Johnston, Ralph McQuarrie, and Mike Pangrazio have a cameo as three Rebels in the final frame; below right: final frame with the hangar ceiling).

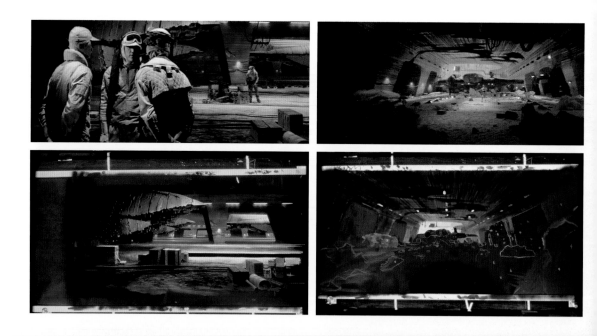

## Echo Base Hangar 2

The enormous Echo Base hangar was shot by the main unit (as directed by Kershner, seated on tauntaun, below right) and second unit from May 22 to June 6, 1979. To make the set seem even more gargantuan, children were dressed as Rebel pilots and placed in the background (below left); using forced perspective, these well-placed extras created the illusion of vast spaces. A limited number of full-sized vehicle models were also sprinkled about and shot so as to give the impression of vast numbers of Rebel Alliance forces (opposite page: full-sized snowspeeders are masked in preparation for final paintings, while a full-sized X-wing faces out with an only-partially built X-Wing behind it; top right: production films Ford as Solo and Daniels as C-3PO).

Unfortunately, during the shooting of the hangar scenes, an ominous note is recorded in the progress report of May 31: "At 7:45 A.M. this morning, John Barry [production designer of the first *Star Wars*] arrived at the studio and complained of feeling unwell. Production arranged for medical assistance and a full report is on file." The legendary production designer died the next day, and the report of June 11 notes that "shooting ceased at lunchtime today to enable any crew members who wished to do so to attend the funeral of the late John Barry."

## Falcon Exterior, Take 2

At the time Lucas made *A New Hope*, there was no way of knowing whether or not the film would be successful, or if he would have a chance to make a sequel. Large sets take up a lot of space and are generally not stored, unless the production is certain that the set will soon be reused. In addition, it actually costs more to break down a set, store it for months or even years, set it back up, and prep it for shooting than it does to simply re-create it. The *Millennium Falcon* exterior set from *A New Hope* was therefore not saved, and a new one was constructed for *The Empire Strikes Back*.

This time the entire ship was built, as it was going to be seen from many more angles than in the first film, and it would appear in three different locations: the Echo Base hangar on Hoth (opposite page and top right), inside the space slug (below right, final frames), and on the Cloud City landing platform (below left)—all three filmed on the *Star Wars* Stage. As was the case in *A New Hope*, the ship was too large to move easily; it was better to leave the *Falcon* where it was first constructed, and have whatever set the script required built around it.

The new ship was not an exact match of the original *Millennium Falcon*. Because the latter would appear in a number of dark locations, new lights were built into its underside. Compare the *Falcon* from *Star Wars* (middle right, with stormtroopers) with the *Falcon* from *Empire* (middle right, with Han running).

## Echo Base Hallways

The stage construct of the Echo Base sets (opposite page, exterior view—the illuminated lights most likely show where production was filming at the time this photo was taken) consisted of the main corridor (which runs through the middle of the frame), the medical center (on the left of the photo), the Bacta tank (note the circular opening on the left), the control room (in the background on the right), and the naturally formed cave where the tauntauns are kept (foreground right). The main corridor appears in the movie to be laser-cut channels in the ice, but were actually wood and plaster (below left, the hallway under construction; right, final frames).

# Echo Base Control Room

The Echo Base control room (below and right), or "ice cave," like the Rebel war room in *A New Hope*, is the nerve center where surveillance and battle operations are coordinated. It was built on Stage 1 and shooting took place in late March. The actors are wearing coats, and though it certainly wasn't freezing, the enormous soundstages were cold enough that the off-camera crew also sported warm clothing (opposite page: sitting on steps, Irvin Kershner talks with Harrison Ford).

## Snowspeeders

Cockpit shots of the snowspeeders (below, final frames) were filmed on an inner-tube rig in front of a bluescreen (opposite page, with Kershner in cap). As was typical for these kinds of scenes, the "sun" light was mounted on a semicircular rail with the ship at its center. The light was able to slide along the rail, thereby creating moving shadows and greatly enhancing the illusion of flight (right).

# Battle of Hoth

As with the tauntauns, the Imperial walkers were mostly photographed complete with their environments at ILM to avoid the compositing problems associated with combining a light foreground with a light background (below, final frame). In most cases, the walkers were standing in front of a large painted backing on a dressed tabletop covered with baking soda to simulate snow (right). The walkers were then animated using the technique of stop-motion. The illusion of movement in stop-motion comes from the animator moving the model a small amount, shooting a frame, moving the model again slightly, shooting another frame, and so forth. With one second of live action consisting of twenty-four separate images, a five-second shot requires 120 manual model-position changes (opposite page: an Imperial walker is about to be taken down by a snowspeeder equipped with a tow cable that is being wrapped around the walker's legs).

## Crashed Snowspeeder

During the attack on the Imperial walkers, Luke Skywalker's snowspeeder is hit by laser fire and crashes (below left, final frame). For this scene, a full-sized "damaged" snowspeeder was built and shipped to Finse, Norway for filming (right: crew in blue parkas examine the craft's cockpit and, opposite page, check the focus range with measuring tape).

Luke just manages to crawl out of the crash-landed speeder before it is stepped on and crushed by an Imperial walker. The crushing of the snowspeeder was accomplished with a two-foot-long miniature at ILM (below right).

# Echo Base Exteriors 1

The Echo Base exterior locations, like other Hoth exteriors, were filmed in Finse, Norway (below right and opposite page). For the most part, they were complete in camera and didn't need to be extended in postproduction. However, the view of the entrance to Echo Base (filmed in part, top right) was completed with a matte painting created at ILM (below left, final frame). Similarly, views of the shield generator were also accomplished using matte paintings (bottom left, final frames).

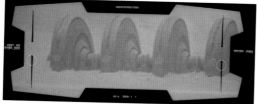

# Echo Base Exteriors 2

A rare panoramic, taken in Norway during the location shoot of the Rebel Alliance's fortified positions at the Battle of Hoth.

## Star Destroyer Bridge

The bridge of Darth Vader's Star Destroyer (the *Executor*)—built on Stage 5 and shot from April 24 to May 2, 1979—was intended to be seen in only a small number of scenes. For this reason, very little of the environment was built as a set: just one instrument bay, the central walkway, the single front window, and one side window alcove (right, in construction). The wide view of the bridge was accomplished by shooting the scene twice, with Imperial officers arranged differently for each shot, then mirror imaging one of the shots to be the opposite side of the bridge. A matte painting filled in the rest of the frame (below, final frame).

This set construction was also used as the bridge of the *Avenger*, Captain Needa's (Michael Culver) Star Destroyer (opposite page, with boom operator in blue and Culver on right); alternate filming of the "two" bridges took place on the same days.

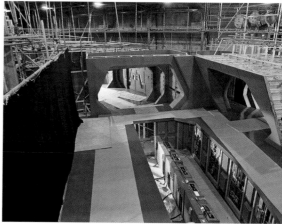

# Star Destroyer Hallway

The Star Destroyer hallway set was yet another flexible utility piece that was used for all the scenes where a nonspecific Star Destroyer interior was needed. Each shot required a distinctly different look, but by emphasizing an alternate direction of a scene, each appears to be a separate portion of the same location. In reality, for example, the control panel bay (opposite page, left; top left, final frame) was across the view screens and hologram table (opposite page, right; top right, a better view; below left, final frame).

# Vader's Meditation Chamber

Darth Vader (David Prowse) maintains a private meditation chamber aboard the *Executor* (opposite page). Built on Stage 5 and filmed from May 2 to May 8, this large room also contains a hologram transmission area where Vader communicates with the Emperor (below right; top left, final frame). The construction was adjacent to the Star Destroyer bridge and hallway sets. The chamber itself included a chair and consoles (below left) and a mechanized arm that holds his helmet (top right).

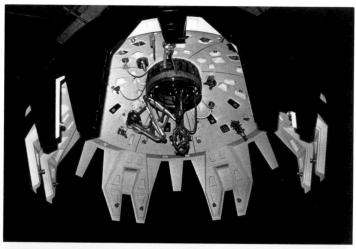

# Millennium Falcon Cockpit, Take 2

Built on Stage 5, this reconstruction of the original *Falcon* is fundamentally the same, although many of its details are different (opposite: actors and crew prepare for a take). Chewbacca's hanging dice have been removed (below right) and the instrument panels along the sides and the back of the cockpit have numerous additional lights and other add-ons (top right), which are evident as Chewbacca (Peter Mayhew), C-3PO, Princess Leia, and Han Solo evade Imperial pursuers (below left).

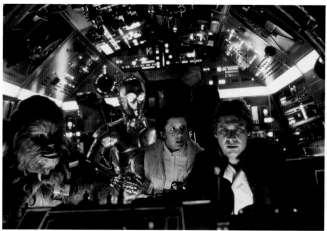

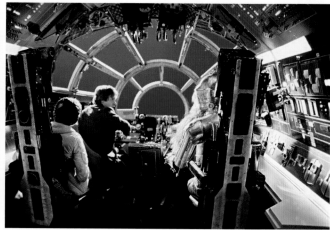

## Millennium Falcon Interior, Take 2

*The Empire Strikes Back* provided an opportunity to show more of the *Falcon*'s interior than had previously been seen in *A New Hope*. The newly designed areas were:

1. A bed where Luke Skywalker recovers after his battle with Darth Vader (below left)
2. A machine room where Han Solo and Princess Leia have their first kiss (opposite page—Fisher is standing in the machine room)
3. A maintenance access area in the floor of the main room where Han tries to fix the hyperdrive (below right)
4. A maintenance access area in the ceiling where Chewbacca works on the ship
5. An elevator that Lando Calrissian (Billy Dee Williams) takes to rescue Luke from the underside of Cloud City (the hatch was built elsewhere). What's driving Lando's elevator? A forklift! (right).

Filming in the "main hold" actually took place on the first day of studio shooting, March 13, 1979, on Stage 8. On May 10, the progress report notes that a small fire occurred in the *Falcon* interior, the "result of a lamp falling over and remaining unnoticed."

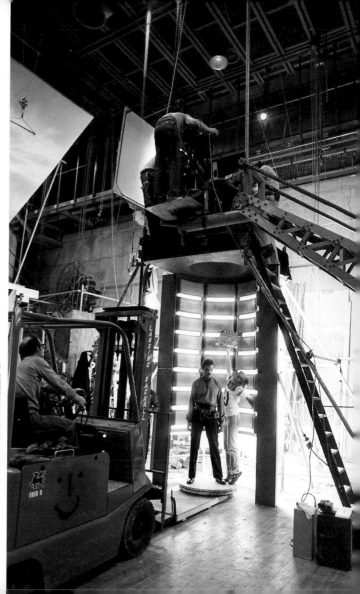

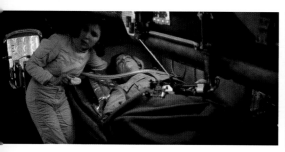

## Asteroid Field

The asteroid field was created thanks to a combination of miniatures and matte paintings. A typical shot included a band of extremely distant asteroids painted on glass, a background asteroid rig (opposite page, bottom left); a layer of mid-ground asteroids on a wire rotation rig (opposite page, top left—Steve Gawley is painting the asteroids; the combination of asteroids can be seen in the two final frames on opposite page, top middle), and a handful of larger asteroid models shot separately in front of a bluescreen. The large asteroid that Han Solo uses as a hiding place for the *Falcon* was a model carved out of foam (below right and opposite page, top and bottom right; final frames utilizing these models are below left and opposite page, middle right).

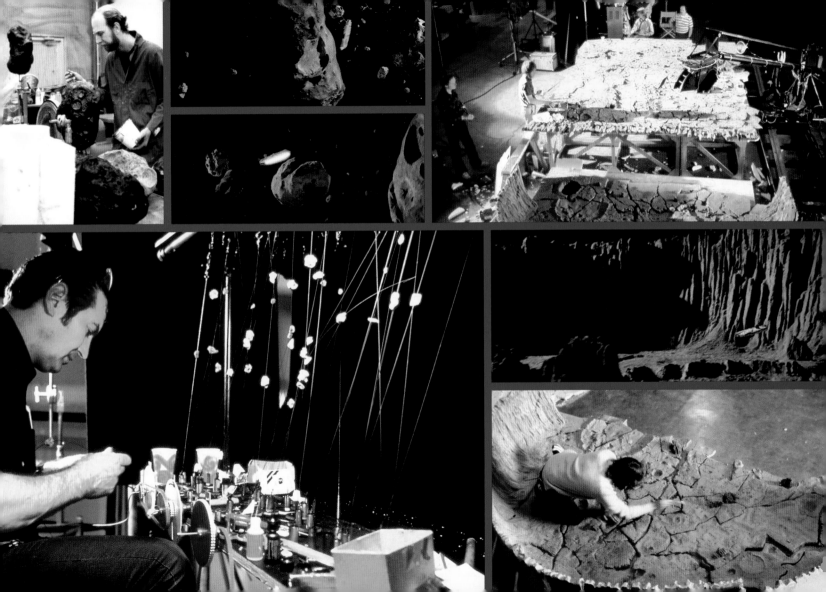

## Space Slug Interior

Undoubtedly the least expensive set built for *The Empire Strikes Back*, the space slug interior was created by laying black Visqueen plastic on the floor of the Echo Base hangar, draping black curtains around the *Millennium Falcon*, and clouding the space with a fog of dry ice (opposite page, with Harrison Ford and a very low-tech mynock). The only wide shot in the sequence was done with a miniature (top left; below left, final frame). The exit from the space slug was also a composite shot including a miniature of the slug's jaws (with Ken Ralston, top right; below right, final frame).

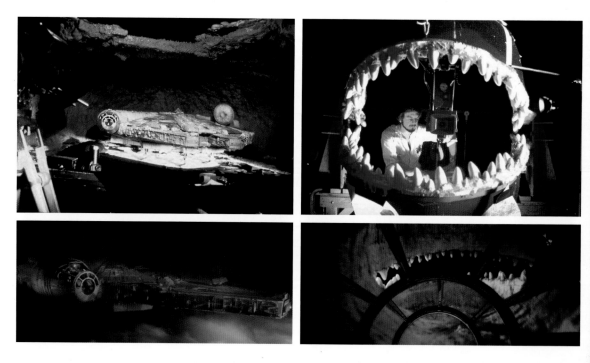

# Dagobah 1

Dagobah, the swampy planet of exiled Jedi Master Yoda (voiced and puppeted by Frank Oz), is primarily a combination of a large indoor set built on the *Star Wars* Stage (opposite page), and a number of matte paintings used for wide shots (below right, with matte painter Chris Evans). The X-wing scenes and much of Luke Skywalker's training was filmed on the soundstage (below left, final frame). Shooting on this set was trying because of the fog used throughout. The crew's first day on Dagobah was August 6, 1979—which was late in a production that was already "forty-five days over" (top right: producer Gary Kurtz, executive producer George Lucas, and director Irvin Kershner). The "Bog Planet" shooting schedule was originally projected for sixteen days of filming, but actually continued until September 21, pretty much the end of the shoot.

## Dagobah 2

The doors to the stage had to be kept closed, and ventilation fans were turned off, in order to maintain the Dagobah set's fog—so the stage became humid and hot (below right: as Luke searches for R2-D2, who's been swallowed). The water in the swamp was not clean, and gradually became more and more foul (top right and opposite page, with Luke on his X-wing). The footing was uneven and sometimes slippery—in fact, the progress reports remark that several crew members slipped and fell into the swamp during shooting, with at least one requiring stitches. Nevertheless, the crew and puppeteers (such as Frank Oz, operating Yoda, below left) soldiered on.

## Yoda's Hut

The exterior of Yoda's humble abode was shot on the same set as the bog on the *Star Wars* Stage (below right), but was dressed to appear like a more distant location. The hovel interior was a separate construct, and was built on Stage 9 (below left, final frame), alongside the tree set—where Luke Skywalker confronts an apparition of Darth Vader. Like the swamp set, the exterior of Yoda's house was built on a raised platform above the stage floor (opposite page: note the seams in the styrofoam; the blocks of styrofoam were carved, then placed together to form the structure) to allow room below for the puppeteers to work (top right: Frank Oz manipulates Yoda, with Lucas smiling); the top of the exterior hovel was also removable .

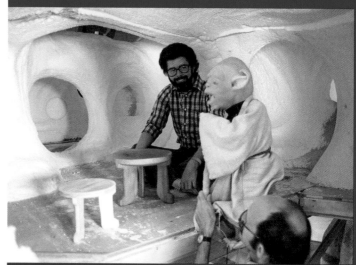

## Cloud City

Cloud City, a floating colony of gas miners above the planet Bespin, was created with a series of matte paintings. The city structure was painted by Ralph McQuarrie on a separate piece of glass (opposite page, bottom) from the cloud background (opposite page, top) to allow the two to move in relation to each other in the shots. To limit how many separate paintings were required, wide shots were used twice. The day shot was mirror imaged, and the second dusk shot was used at a different size to vary it from the first (below).

 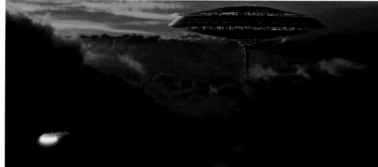

# Cloud City Landing Platform

The Cloud City landing platform (right, final frames) existed in three different forms: on the *Star Wars* Stage, as a full-sized floor under the *Falcon* set that the characters would walk on (below right); as a six-foot long miniature that the *Falcon* model lands on (under construction, below middle); and in a wide shot where matte paintings extended the set (below left, with Ralph McQuarrie).

     To complete the landing platform shot, matte photographer Neil Krepela used front projection (opposite page, with Craig Baron in background). The plate image was projected through a beam splitter (a half-silvered mirror) so that the camera (which is photographing the painting combined with the live-action plate) could effectively be in the same place as the projector, thus eliminating perspective distortion.

# Cloud City Interiors 1

In an early maquette of Cloud City (opposite page), it's obvious that a limited number of sets—a straight and curved section of corridor; a bedroom; an unidentified locale; and the dining room where Han, Leia, and Chewbacca are ambushed by Darth Vader—were multiplied into about twenty seemingly different sets. Situated in Stage 2, and shot on from April 5 to April 24, 1979, the Cloud City interior sets were for the most part closed in and windowless (below left and middle). In a handful of cases, a painted backing (below right), or a full matte painting, were used to provide glimpses of the outside.

## Cloud City Interiors 2

As in the Rebel blockade runner and the Death Star corridors, characters in Cloud City can be seen exiting from one angle and then, following a cut, reentering the same set from another angle. Occasionally, the sets would be repainted—as in the case of the red-brown corridor outside the prison area (opposite page), which used to be white (right).

## Carbon Freezing Chamber

Located on Stage 4, the Cloud City carbon freezing chamber is where Han Solo and Princess Leia declare their love for each other, and where Luke Skywalker fights Darth Vader. The original design for this set was much more compact (below right: maquette), but the look was altered to be roomier in order to accommodate the fight choreography (below left, under construction; opposite page, the set). Barely visible in the finished film is a second freezing chamber on a painted backing (top right, in background).

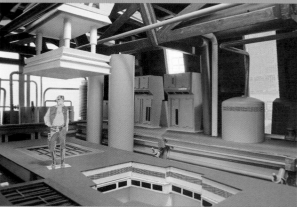

# 76

## Gas Processing Vane 1

Shooting on the gas processing vane built on Stage 1 and referred to as the "gantry platform and pinnacle," took place between July 26 and August 3, 1979. The gas processing vane set made ingenious use of two painted backings—an up/down backing and a straight-across backing (right, both in midconstruction—the up/down backing is on the floor; below, the up/down backing complete and in place at a forty-five-degree angle)—in addition to a number of matte paintings, to suggest the thousand-foot-diameter interior of the vent shaft (opposite page, without either backing).

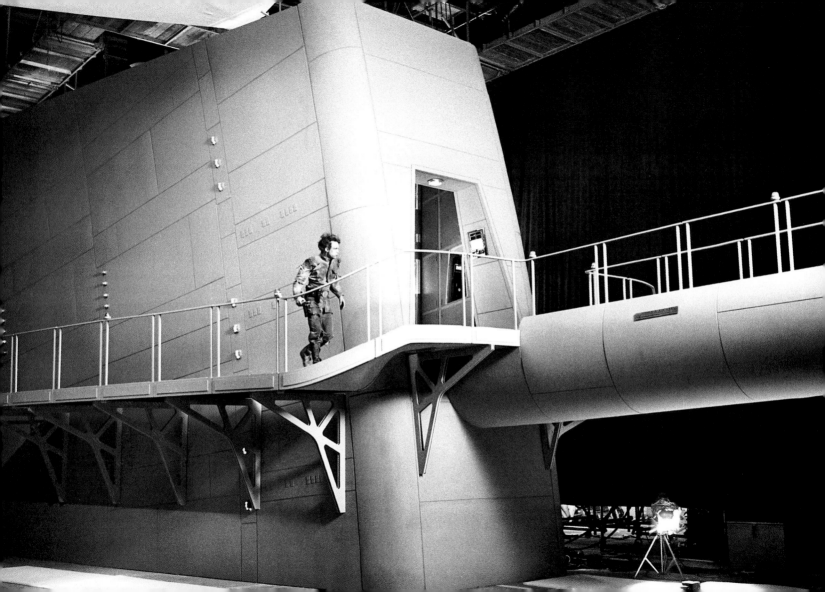

## Gas Processing Vane 2

To avoid this lengthy sequence becoming a vast bluescreen compositing nightmare (because of the difficulty of getting good extractions from Darth Vader's shiny black helmet, and because of the smoke and sparks from the duel), most of the sequence was shot in front of the two backings, both of which were flat, but painted with built-in perspectives of the inside of the cylinder. One backing was created for up/down angles (opposite page, lower-left corner); the other was painted with a straight-across perspective (opposite page, behind the scaffolding on the right).

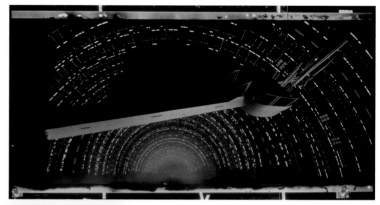

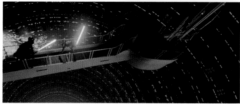

Matte paintings were used to create wider shots. A final frame (left) combines the matte painting (above) with the plate shot during principal photography (opposite page; the camera is at the top of the scaffolding next to the two lights).

The up/down painted backing behind the set; note how the smoke and sparks blend perfectly over the background, since there is no compositing.

The same up/down backing now rotated 180 degrees and suspended over the set.

The straight-across perspective backing.

## Cloud City Underside

After being sucked into a debris trap and unceremoniously dropped out of a hatch on the underside of Cloud City, Luke Skywalker manages to grab hold of an antenna and hang on until he's rescued by Princess Leia and Lando Calrissian in the *Millennium Falcon* (below, final frames, completed with a matte painting; bottom strip, a different discarded version of the rescue in which Lando drags Luke across the surface of the *Millennium Falcon* into the hatch).

Built on Stage 2, the underside of Cloud City existed in two forms: a full-sized set of the antenna and hatch (opposite page), and a forced-perspective-arc model used for vista views (right, Nilo Rodis in blue shirt). A number of the shots of the underside were taken with moving cameras, so a miniature was chosen over a matte painting to enable true perspective shifting.

## Medical Frigate

The end of *The Empire Strikes Back* takes place aboard the medical frigate set—built on Stage 8 and shot on July 11 and July 12, 1979—where Luke Skywalker is fitted with a mechanical hand (top left: final frame; opposite page: Anthony Daniels, Mark Hamill, Carrie Fisher, producer Gary Kurtz, director Irvin Kershner, pointing). The last shots of the film are of Luke, Leia, R2-D2 (Kenny Baker), and C-3PO looking out through the frigate window (top right), and then a close-up of the droids (below right, being filmed; below left, final frame). The camera then pulls back to reveal the Rebel fleet heading away. The footage of the medical center was rear-projected onto a window in a miniature of the frigate.

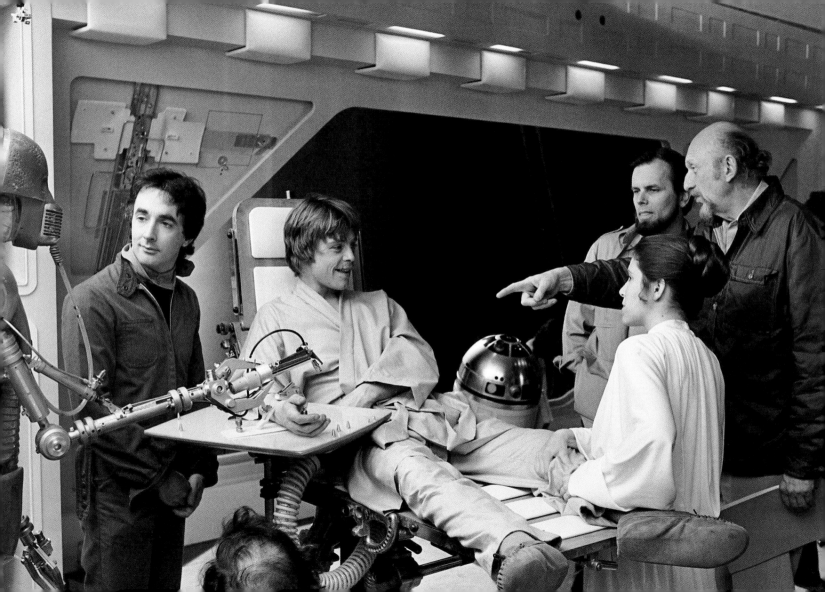

STAR WARS: EPISODE VI

COMING
OF AGE

## Expansion

The conclusion of George Lucas's *Star Wars* trilogy marked the return of the seasoned crews of *A New Hope* and *The Empire Strikes Back*—veterans such as Norman Reynolds, Ken Ralston, Steve Gawley, Lorne Peterson, Joe Johnston, Dennis Muren, Ralph McQuarrie, and Richard Edlund. Everyone was ready to do something even more ambitious than before. Consequently, the sets and the number of special-effects shots and their complexity reflect Lucas's greater vision—especially in terms of what ILM was asked to do (opposite: the *Return of the Jedi* model shop, as it sets up for a studio shot with George—see page v; note the many models and maquettes on the tables, including a creature from the 1982 film *Poltergeist* hanging from the ceiling).

Principal photography began on Monday, January 11, 1982, once again at Elstree Studios in London, where scenes were filmed until Thursday, April 1. After that, production moved to Yuma, Arizona, where shooting continued from Monday, April 12 to Saturday, April 24 (second unit shot on April 26 and 27); everyone then continued to Crescent City, California, where many of the Endor scenes were filmed from Monday, April 26 to Saturday, May 8 (second unit continued until Saturday, May 29). Some first-unit footage was then recorded at ILM in San Rafael from Monday, May 10 to Monday, May 17. One of the last shots on record was stuntman Bob Yerkes being catapulted into the desert, and was performed in Northridge, California, on January 30, 1983.

*Return of the Jedi* (1984)

EXECUTIVE PRODUCER: George Lucas
DIRECTOR: Richard Marquand
PRODUCER: Howard Kazanjian
STORY: George Lucas
SCREENPLAY: Lawrence Kasdan, George Lucas
PRODUCTION DESIGNER: Norman Reynolds
DIRECTOR OF PHOTOGRAPHY: Alan Hume
EDITORS: Sean Barton, Marcia Lucas, Duwayne Dunham
ART DIRECTOR, VISUAL EFFECTS: Joe Johnston
VISUAL EFFECTS: Richard Edlund, Dennis Muren, Ken Ralston
CONCEPTUAL ARTIST: Ralph McQuarrie
ART DIRECTORS: Fred Hole, James Schoppe
SET DECORATORS: Michael Ford, Harry Lange
MAKEUP AND CREATURE DESIGN: Phil Tippett, Stuart Freeborn
MATTE PAINTING SUPERVISOR: Michael Pangrazio

## The Second Death Star

The opening shot in *Return of the Jedi* reveals a second Death Star under construction in orbit around the moon of Endor. Like its forerunner in *A New Hope*, the Death Star 2 was a composite of three versions: two miniatures, one when seen from a great distance (right) and one when seen up close (below); and matte paintings (opposite, with Chris Evans), when seen from a medium distance.

# Death Star 2 Hangar

The hangar or "main docking bay" set, built on Stage 6 (opposite and below), is seen three times: at the outset of the film when Darth Vader (David Prowse) arrives; upon Emperor Palpatine's (Ian McDiarmid) dramatic entry (right); and during the climactic death of Anakin Skywalker. Though this set was very large, it still represented only a fraction of the whole hangar, and matte paintings were used to expand the environment.

Incredible secrecy was maintained in relation to the reveal of Darth Vader's human visage beneath his mask. The elder Skywalker was played by Sebastian Shaw, but the shoot's progress reports refer to him only as "The Man." His brief but noteworthy appearance in the hangar was filmed on Wednesday, February 24, 1982, while his after-death apparition was recorded against black velvet on Stage 5 on March 2, 1982.

## Jabba's Palace, Exterior

The exterior of Jabba the Hutt's palace existed as a series of matte paintings and as a full-size door set, which was meant to be about forty feet across (opposite). Because most of the scene with the principal actors would be played in relatively close shots, it was less expensive to limit the scope of the build to include just enough set to cover these close-ups, and extend the wide establishing shot as a matte painting (below left, final frame).

The scene of C-3PO (Anthony Daniels) and R2-D2 (Kenny Baker) as they approach the palace door was filmed in Death Valley's Twenty Mule Team Canyon on December 11, 1982.

A wide shot that shows the transition from day to night went through several iterations: a tabletop miniature was created for the puppet to sit on, which was then photographed in front of a painted backing (below right, with Phil Tippett, standing, and Ed Hirsh); another version has the frog fairly large in the foreground (top right); the final version has the frog in the lower-left corner of the frame (top left, final frame).

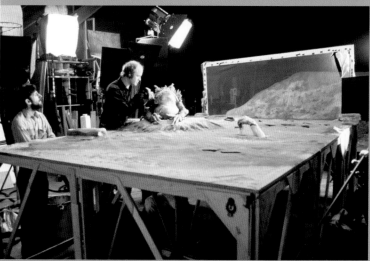

## Jabba's Throne Room

Because this set was filled with puppets that would need to be operated from underneath, Jabba the Hutt's throne room was built about eight feet above the floor of Stage 8, which actually worked well within the reality of the film, since his throne slides back to reveal a trap door (below right, being constructed; below left, during a take; opposite, the droids are on boxes, perhaps to give the camera operator a better angle, or perhaps because the floor grating posed potential problems for R2-D2's wheels). Theoretically, the puppeteers were standing in Jabba's pit, and the additional space gave them adequate room to work. The palace hallway was constructed as an interior that also included the other side of Jabba's palace gate. Because these throne room scenes were being shot in late January 1982, the middle of winter, nearly the entire main-unit camera crew came down with the flu; Lucas himself came down with influenza on February 1.

Later, in postproduction, Dave Carson wore a Jabba hand (right, with Phil Tippett) for an insert shot of the Hutt reaching into a bowl to grab something to eat.

## Rancor Monster Pit

Below Jabba's throne room, the Hutt gangster keeps a pit with an enormous monster—the rancor—and delights in the horrible spectacle of watching his victims being eaten alive. The rancor was an approximately two-feet-tall puppet (right, with Kim Marks). The pit was created in two sizes: one at ILM scaled to match the puppet (below left, with Dennis Muren on right), and a full-size one built on Stage 1 (below right and opposite, with Mark Hamill held by a life-size rancor paw, and director Richard Marquand holding a bone), with the adjoining dungeon cell corridor built on Stage 7. Most of the shots in the sequence were filmed combining both of these sets, with skillful matching and editing tying them together visually; additional rancor pit shooting took place on Elstree's Stage 9 against a blue backing.

# Jabba's Sail Barge Exterior 1

Three methods were combined to create Jabba the Hutt's sail barge: a detailed miniature for motion-control photography (below left, with Richard Edlund; top right, with Don Dow); a breakaway pyrotechnics model, for the explosion of the barge; and a full-size set (opposite). The latter was built in Buttercup Valley near Yuma, Arizona, on a flat section of desert surrounded by sand dunes (below right), and filmed in April 1982.

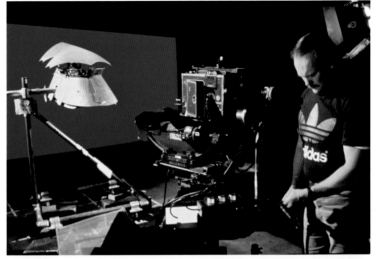

## Jabba's Sail Barge Exterior 2

The sail barge (seen here under construction) makes it clear that they built the Great Pit of Carkoon from the ground up instead of digging down, and then filled in the area around the barge to complete the illusion.

 **89**

# The Great Pit of Carkoon

Jabba decides to execute Luke Skywalker (Mark Hamill) and Han Solo (Harrison Ford) by throwing them into the mouth of the Sarlacc. The monster's pit was created in a surprising manner: A twenty-five-foot-high platform was built up from the desert floor in Yuma, Arizona, with a sand-covered cone indented into it (right) rather than digging into the ground. The original model planned for a fifty-foot-high platform (below; in the "pit" are Norman Reynolds, George Lucas, and Richard Marquand; on the left of this architectural model is a miniature stairway, which looks like it goes up four floors!); in the end, however, it was decided that the build was just too massive, and the height was reduced by half.

The sand skiff was supported by two strong posts (opposite). Clever camera angles and time of day were carefully chosen to hide the supports and their shadows, creating the illusion that the skiff was hovering. Miniature work was completed at ILM.

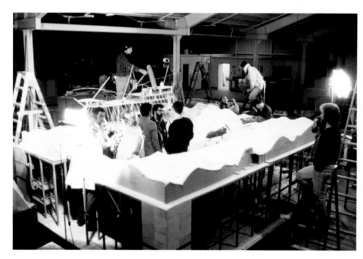

## Jabba's Sail Barge, Interior

The sail barge interior was built on Stage 9 (below, with a dustbin from Stage 6). The set's venetian blinds helped create continuity with the desert exteriors shot in Tunisia; the effect was created by allowing light generated by electric lamps to overexpose the film as it filtered though the slats, giving the impression of intense heat (opposite). This set also had a roof, to reinforce the claustrophobic feeling of the scene and Princess Leia's (Carrie Fisher) enslavement.

## Sandstorm 1

After escaping from Jabba's sail barge and the Great Pit of Carkoon, Luke Skywalker, Princess Leia, Lando Calrissian (Billy Dee Williams), and Han Solo head back to their ships parked in the desert. A sandstorm closes in around them as they approach the vehicles (opposite and right). This scene was the only one in *Return of the Jedi* requiring a *Millennium Falcon* exterior. The size of the *Falcon* set built for *The Empire Strikes Back* meant that it was too large to save, therefore it was rebuilt for *Jedi* on Stage 2. However, because the sandstorm would greatly limit visibility, only half of the *Falcon* was reconstructed (opposite and below). During postproduction, Lucas decided that this scene—shot on the first day of principal photography—was unnecessary, and deleted it from the picture.

## Sandstorm 2

This intriguing matte painting shows the *Millennium Falcon* and Luke's X-wing parked on Tatooine beneath a sunny sky. The black hole (below) would have been filled in during postproduction with an insert of the actors. Perhaps as they descended into the canyon, the sandstorm hit.

# Dagobah, Take 2

Once again, a set too large for storage had to be reconstructed. Luke Skywalker's return to Dagobah to complete his training with Yoda (voiced and puppeteered by Frank Oz, with the aid of his crew: Mike Quinn, David Barclay, and David Greenaway) required the re-creation of part of the bog planet on Stage 2 (opposite and below: final frames show reverse angles covering the whole set), as well as the interior of the aged Jedi Master's abode on Stage 1. For a handful of the exterior scenes, Alec Guinness reprised his role as Obi-Wan Kenobi on Wednesday and Thursday, March 10 and 11, 1982. He also performed his final appearance (for the Ewok village scene, filmed in front of black velvet) on that Thursday.

## Mon Calamari Cruiser Briefing Room

Rebel forces plan their attack on the second Death Star aboard the Mon Calamari cruiser, referred to at the time as the "H.Q. Frigate" (opposite, with Ford, Fisher, Hamill, and Peter Mayhew—without Chewbacca mask—in foreground, and director Richard Marquand gesturing next to table). Built on Stage 5 (below right, under construction), this set is actually two in one: The briefing room is also the control center where Admiral Ackbar coordinates the attack on the Death Star (below left: Ackbar, played by Tim Rose, is in white; note the circular graphics displays common to both the briefing room and the control center).

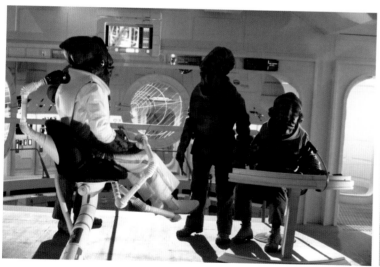

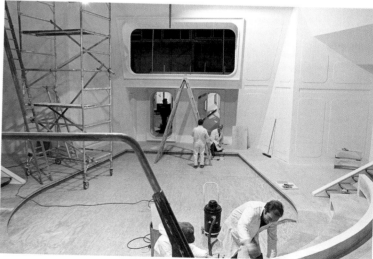

## 95

## Mon Calamari Cruiser Hangar

The hangar of the Mon Calamari Cruiser consisted of a floor, a small section of wall, a repainted Imperial shuttle, and one and a half X-wings (opposite, with the X-wing half on the right). The scene with Han Solo and Lando Calrissian was mostly framed against these set pieces. While they discuss their fate and the fate of the *Millennium Falcon*, that starship is visible in the background of a number of shots—ironically, the full-size *Falcon* exterior built for the sandstorm couldn't be used here. Scheduling prevented this scene from being filmed on the same stage as the *Falcon* exterior, so matte paintings were used instead (top right, with Michael Pangrazio working on the establishing shot; below right, for coverage of Lando and Han talking).

## 96

## Death Star Throne Room

Built on Stage 4 of Elstree Studios, Emperor Palpatine's throne room was a big, complicated set—and possessed another one of the *Star Wars* saga's bottomless pits (below left: an early art department model). There was a little bridge over this particular pit, complete with safety railings (top and below right: note that the pit isn't exactly bottomless). One of the railings was constructed so that when it's hit during Luke and Vader's climactic battle, it would seem to be split by a lightsaber blow (opposite: Hamill as Luke and Prowse as Vader, with Richard Marquand standing behind camera). The catwalks were part of this construction, though they're not seen here.

The Emperor's throne room scenes were filmed from Thursday, February 25 to Thursday, March 4, and from Friday, March 12 to Tuesday, March 16, 1982 (second unit continued shooting on the set for a few days after that).

## Imperial Shuttle Cockpit

Han Solo's strike force—Chewbacca, Leia, C-3PO, and Luke—uses a stolen enemy shuttle to sneak past Imperial forces and land on the forest moon of Endor (below left). The rear portion of the shuttle, behind the cockpit, was designed as a troop carrier but was never seen in the film (opposite, filled with crew members: because Lucas wasn't in England during much of preproduction, photos like these were taken and sent to him for set approvals). Of course, the shuttle cockpit didn't look out into space—it had a view of a wooden wall (below right).

## Star Destroyer Bridge

The Star Destroyer bridge, built on Stage 1 at Elstree Studios for *Return of the Jedi*, was far more complete than the much smaller set created for *The Empire Strikes Back* (opposite, during a rehearsal in which Vader is a stand-in with only mask and cape—notice the ungloved hand—and a black droid is posed on the left). Because the production art department lobbied for it, and given the increased number of scenes on the bridge in *Jedi*, Lucas decided to have more of the set built for this final installment (below, with Prowse as Vader and Kenneth Colley as Admiral Piett).

## The Forest Moon of Endor

The exteriors for the Endor forest scenes (below, final frames) were shot in Crescent City, California (opposite, with Anthony Daniels in C-3PO costume but without mask, as crew and extras look on). The various Ewoks and crew and extras were put up in motels scattered around the area—Smith River, Thunderbird, Bonn, Pacific, Ship-A-Shore, Travel Lodge, Northwoods. One day, then-production-assistant Ian Bryce, whose job was to round up the Ewoks for each day's shooting, couldn't locate them. Frantic, he went in search of the strays—only to find the actors back on location, where they'd been hiding. They finally appeared wearing "Revenge of the Ewoks" T-shirts.

# 100

## Speeder Bike Chase

The speeder bike chase sequence was all visual effects, from its blazing start to its explosive end. Foregrounds were shot bluescreen at ILM, either with the actors on full-size speeder bike props (below, with Fisher as Leia on the main stage) or with rod-operated puppets (opposite, on a model mover being filmed by Mike McAlister). These were among the last shots to be filmed during principal photography in May 1982.

Backgrounds were a challenge, because the speeders were meant to look as though they were flying along at over 100 miles per hour through closely spaced giant trees. The narrow clearances and uneven terrain precluded the use of a helicopter, a camera car, or a dolly on a track. The solution, however, was brilliant: Have the cameraman use a Steadicam, which was only then coming into mainstream use, and simply walk through the woods with the camera running at one frame per second (right, with the inventor of the Steadicam, Garrett Brown). A taut length of fishing line strung between trees along the preset path provided the operator with a straight reference line to follow (this string was too thin to be seen on camera, and didn't show up in the shots). When the film was later run at normal speed—twenty-four frames per second—it seemed as though the camera was careening along at tremendous speeds.

## Ewok Village

The Ewoks live high above the forest floor in structures attached to the trunks of large trees. The Ewok Village was built inside Elstree Studios' Stage 3 (below, under construction), where shooting took place from Thursday, January 14 to Monday, February 1, 1982 with both the first and second units. A painted forest backing hung behind the tree trunks, platforms, and huts (right, behind balcony). Wide shots were created using matte paintings (opposite, with Chris Evans). More second-unit footage was filmed at ILM and Skywalker Ranch with Mark Hamill and some of the Ewok extras on Monday, December 20, 1982. The last second-unit pickup of the "bonfire" scene was needed the night of Sunday, February 27, 1983—and was shot at 9:30 P.M. on the ranch with just Hamill on camera.

## Imperial Landing Platform

It's here that an Imperial officer (Bill Hoyland) hands over Luke Skywalker to Darth Vader (below left, final frames)—now revealed to be father and son. The Imperial landing platform on Stage 3 featured a partial exterior of an AT-AT walker along with one "tree" (below right and opposite: art department models of set with cutout of Darth Vader). To the left one can see the head of the transport, which is a flat, painted backing—the perspective, however, is off and in actuality represents an impossible angle.

## Shield Generator Bunker, Exterior

Like the other Endor scenes, the exterior of the shield generator control bunker was built in Crescent City, California (below: as they film Solo protecting Leia, a crew member uses a sheet of Plexiglas to protect the camera crew from on-set pyrotechnics). Once again, editing creates the illusion that anyone entering through the exterior door in California would arrive at the set's interior—built in London, on Stage 8 (see spread 104). When the Rebel Alliance attacks the bunker, a battle between stormtroopers, Rebels, and Ewoks ensues—preparations for which involved additional dozens of cast and crew (opposite: an assistant director explains the logistics; note the ladder leading up to an AT-ST, or "chicken" walker, on left).

# 104

## Shield Generator Bunker, Interior

The interior set at Elstree Studios was partially complete, and included the bunker corridor (below right, with Ford, Fisher, and extras). One side of the set featured a bluescreen (opposite), where a miniature of a huge power generator was added later (below left, final frame).

## AT-ST Interior

During the battle of Endor, Chewbacca and two Ewoks manage to capture a "chicken" walker, or AT-ST (below left, the cockpit chairs seen from behind). The two walker crewmen who are subdued were played by co-producers Jim Bloom and Robert Watts (below right). Visiting the set that day, George also subdued one of the Imperials (opposite, with Bloom).

## 106

## Battle of Endor 1

This sequence was partially shot as miniatures at ILM, right next to the door of the main stage of the "C" Building, where for many years Javva the Hutt (the makeshift coffee bar of ILM) stood. One of the AT-STs was shot as a miniature in a model forest (opposite, with Paul Huston and Larry Tan; top right, final frame). Wily Ewoks let loose an avalanche of felled trees, which become entangled in the AT-ST's legs, provoking its explosion—shot at high speed (below left, at ILM; below right, final frame).

## 107

## Battle of Endor 2

Life-size and miniature "chicken" walkers meet with various fates as the tide of the climactic battle turns:

Dennis Muren, on ILM's motion-control stage, prepares the miniature AT-ST for . . .

. . . a sequence that melds the model with a plate shot in Crescent City, where the AT-ST moves aggressively through the Endor forest blasting Ewoks . . .

. . . but it actually walks into a trap—wherein two giant logs are released and collide on the machine's head. For the explosion, the miniature was rigged with pyrotechnics at ILM.

A live-action plate of the AT-ST, moments before Chewbacca, on location, climbs into the cockpit of the AT-ST—which was mounted on the head of an earthmover (opposite).

## A-Wing and B-Wing

In space above Endor, the Rebel Alliance is attacking the second Death Star. A large part of the Rebel fleet is composed of A-wings and B-wings. For the former, a full-size cockpit was built in London (top right, final frame), but the latter was seen only from the exterior, so no cockpit was necessary. Both starfighters came to life through model making. Typically you build one pattern, and then mold it to create a cast to make copies (below right: Wesley Seeds prepping the A-wing model to go into the mold), which are then painted (below left: Seeds painting an A-wing model). But even if you're going to make only one you have to go through this process. Because the pattern is often carved out of wood, a fiberglass shell is created so there's room for an armature (opposite, Scott Farrar shooting the B-wing model on the motion-control stage).

## The Gunport That Never Was

In addition to building a *Millennium Falcon* that didn't make the final cut, production re-created the ship's gunport (opposite, a painter at work; top right, a front view of the gunport set; below right, a side view of the *Falcon* corridor exterior). A maquette (below left and middle) also makes plain what some fans have always wondered about: Where is the medbed, on which Luke rests at the end of *The Empire Strikes Back*, in relation to the rest of the interior? The answer is that it's just off the main room where the hologram table is (below left, room at top left of image).

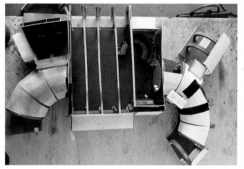

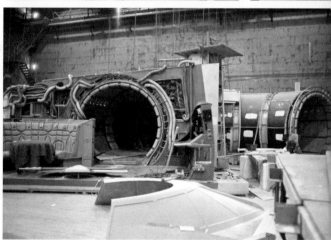

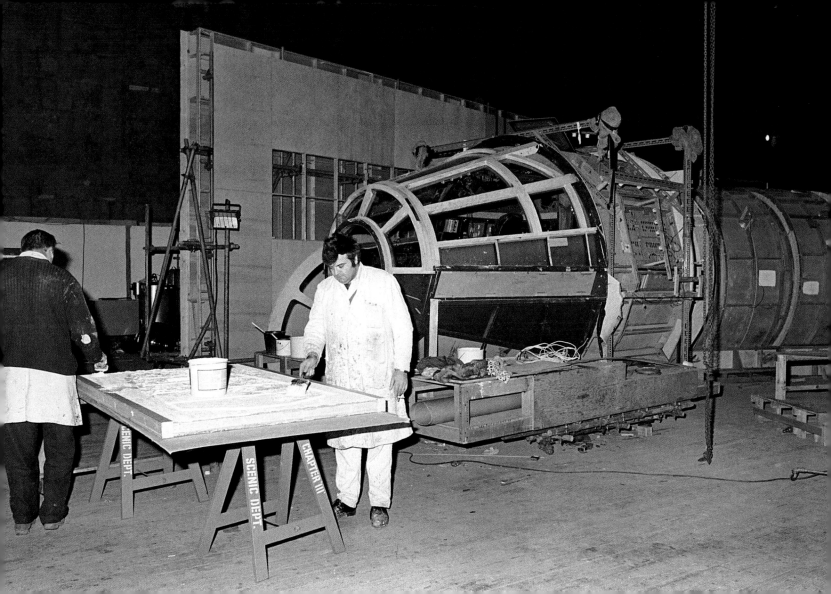

## Death Star Surface

At ILM, a section of the Death Star exterior was built for a series of shots in which the Rebels fly up to the superstructure and dive in through a circular opening (opposite, with Gene Whiteman and Richard Edlund; the aperture is on the right). This area is the same circular opening they burst out of after their breakneck chase through the Death Star's interior; a tunnel section was also built into this portion. This particular miniature is set up vertically because the camera would otherwise have to point straight down—and the camera being used, the VistaCruiser, could just about point straight down, but not quite. Building it horizontally would have limited the depth to about four feet. Nevertheless, for shots that weren't pointing straight down, a separate, horizontal miniature was created for when the Rebel ships are swerving around the gun towers (right, miniature; below, final frames).

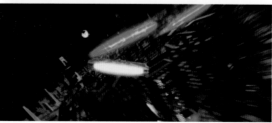

## Death Star Interior

More miniature setups and painstaking camerawork at ILM were required for the desperate pursuit down narrow corridors to the Death Star's central reactor. Cockpit shots of Lando Calrissian in the *Millennium Falcon* (filmed separately on a stage in Elstree Studios) were added later. One of the Death Star interior tunnels was built with removable ceiling pieces, parallel to the motion-control tracks (opposite; and final frame below, with TIE fighter). As the camera moves along at one frame per second, portions of the ceiling are yanked out of the camera's way; but the illusion is one of an enclosed space. Another tunnel was constructed for pyrotechnics (right). The Empire High-Speed Camera rocketed down this area while an explosion "chased" it—all of this was filmed at seventy-two frames per second (below: final frame with explosion and the *Falcon*).

# 112

## Death Star Reactor

The previously unseen central reactor of the Death Star (opposite, with model-shop supervisor Steve Gawley) is revealed in *Jedi* just as the *Millennium Falcon* and an X-wing locate its core and obliterate it. This impressive chamber, miles across, was built as a sixteen-foot-wide miniature (opposite). The blue energy effects around the center were shot separately. A laser aimed at a rapidly spinning mirror emitted a cone of light. When passing through clear air, the beam of the laser wasn't visible. Smoke blown through the ray, however, was brightly illuminated and created the energy effect (below, final frames).

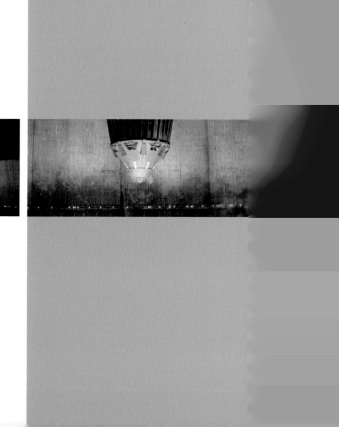

STAR WARS TRILOGY SPECIAL EDITION

# OFF THE SHELF

## A New Hope Special Edition

From 1993 to 1994, I experimented with the idea of using off-the-shelf software to create computer-generated (CG) final shots. I made an X-wing CG model and some TIE fighter models, and created a four-second-long dogfight. It was a fun shot but it was also a technology test: Can one do stuff like this with off-the-shelf software? I thought it came out pretty well and eventually showed it to George. I really didn't know him all that well then, and he just said, "Yeah. Hmm," and that was the end of that.

At the time, I had some strong opinions about how the ILM computer graphics department should operate. It had been built up around the idea of doing really complex creature work. *Terminator 2: Judgment Day* (1991) and *Jurassic Park* (1993) had been the big landmark films that we were famous for—but we didn't have any good way of doing some of the things that inexpensive commercial systems could do. We just weren't geared toward creating simpler hard-surface shots (vehicles, metal textures, etcetera) since we didn't have trained hard-surface modelers. Everything went through the creature pipeline instead, which was great for doing these really complex things that nobody else could do, but we couldn't compete when creating simpler objects. Consequently, the numbers we bid for hard-surface CG shots were astronomical (like all visual effects houses, ILM makes bids for films; depending on a variety of factors, we then either get the assignment or not).

When we were starting up on the *Special Edition* of *Star Wars*, visual effects producer Tom Kennedy already had pinned to his office wall a whole bunch of space-battle storyboards from *A New Hope* that George wanted to redo. So I said, "Tom, I want to take a crack at a couple of these." He was a little skeptical, and decided to do a parallel test. We picked two shots that were technically equivalent—I would do one of them my way, and the computer graphics department would do the other one their way. I spent a little over a day animating mine, another day lighting it, and another day running the renders and putting the composite together. Meanwhile, the test over in computer graphics went on for about two and a half weeks; they were making slow progress and eventually the plug was pulled. (Top right, my test footage from 1993; middle right, the CG department's attempt; bottom right, a quick preview render for one of the *Special Edition*'s new shots, known as "SB1"; opposite, a grid of *Special Edition* final frames.)

The next step was to show George. He was pretty happy with it, so I got the contract. "Okay," he said. "You want all these shots? They're yours."

*Star Wars Trilogy Special Edition* (1997)

DIRECTOR: George Lucas
PRODUCER: Rick McCallum
VISUAL EFFECTS SUPERVISORS: Alex Seiden, John Knoll, Dave Carson, Stephen Williams, Dennis Muren, Joseph Letteri, Bruce Nicholson
EDITOR: T. M. Christopher

## The Empire Strikes Back Special Edition

I had just finished the final "final" on *The Empire Strikes Back Special Edition*. It was a Friday
(in January 1997) and I was at ILM's wrap party for *A New Hope Special Edition*. There
were tents in the courtyard, everyone was having a good time, and I ran into George, who
said—and this was exactly six days before we were supposed to ship the last CG film out for
*Empire*—"I have one more shot I want to do for *Empire*. But we can talk about it Monday."

And I said, "No, no—we have to talk about this now! What is it?"

"After Han's *Falcon* is parked on the Star Destroyer," he said, "it's not clear that Boba
Fett is in the junk, too, and that he's following Han. I think we need one more shot in there
to explain what's going on."

So that night I went home and got out my *Art of The Empire Strikes Back* book and used
it as reference for a really quick CG model of Boba Fett's ship, *Slave I*. I paired that with an
already constructed CG *Millennium Falcon* model.

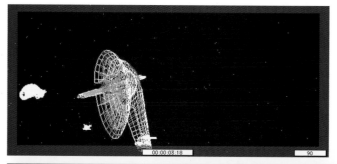

Over the weekend I cobbled together an animatic (a three-dimensional digital storyboard)
of *Slave I* clandestinely following the *Falcon*, and showed it to George on Monday (top right,
a wire-frame of BF1—the "Boba Fett one" shot). He had a couple of comments on the timing,
so we started a revision. Meanwhile, Rod Woodall had been working, as a personal project,
on building a detailed CG model of *Slave I*, so I got him to give me a copy with the textures.
The rest of that Monday we worked on the shot revision, also completing the CG version of
*Slave I*. On Tuesday morning, I showed George the revision of the animatic, and he approved
it. The afternoon was spent lighting the shot, and by the end of that evening I was pretty
much satisfied. I spent Wednesday continuing to light and render elements. Thursday, I put
the comp together, and filmed it Thursday night. We screened the completed version Friday
morning, and George gave it the final thumbs-up (bottom right, final frame). If anything had
gone wrong during that week, we would've missed the deadline. (Other sequences were also
enhanced in *The Empire Strikes Back Special Edition*. Opposite, Howie Weed in the wampa suit.)

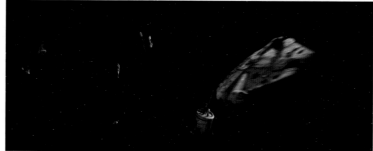

## Return of the Jedi Special Edition

The Coruscant celebration shot, which appears at the very end of the *Special Edition* version of *Return of the Jedi*, was something I really wanted to do. For a long time there had been talk about the conceptualization of this particular planet, which is one big city. In fact, Ralph McQuarrie had done paintings of it quite a while back. Coruscant promised to be really different from anything that had been depicted in *Star Wars* before.

The final shot was a mixture of elements. Some of the Coruscant buildings—three or four—were built as miniatures, and Brian Flora did the matte paintings. The whole plaza had to be populated with extras, so we did a live-action shoot on the main stage. I designed the camera movement, and got approval from George. We used one of our track cameras switched over to a servos, so we could really haul down the track. We had scaffolding built and put the heavy motion-control dolly track on top of it, about twelve feet off the floor, which meant we had this three-thousand-pound camera just careening down the track—it was a big, scary thing to watch. By the time we got to the mid-point of the shot, the camera was probably going over fifteen miles per hour. But that was necessary because it was a sixty-foot move that covered a lot of ground. It was the longest and most complicated shot I did on the three *Special Editions*; creating and putting all the elements together took about a month. (Right, final frames; opposite, the scene in Jabba's throne room was also enhanced in the *Return of the Jedi Special Edition*.)

STAR WARS: EPISODE I

# A NEW BEGINNING

## VFX Supervisor

As it turned out, volunteering to do the space battle shots for the *Special Edition* of *A New Hope* was a good idea. I got to know George Lucas, and once he knows you pretty well, it becomes more and more casual. In fact, about halfway through the work on *Special Edition*, Jim Morris (former president of ILM, now a producer at Pixar) pitched the idea to George that I should be one of the visual-effects supervisors on Episode I. George said fine, so I was slated as the first one to start, during principal photography, with supervisors Dennis Muren and Scott Squires coming on later as we ramped up.

Principal photography began at Leavesden Studios, England, on Thursday, June 26, 1997. (Opposite: battle droid mannequins on a soundstage. Early on we had concerns about how we were going to accomplish the effects of lightsabers slicing and dicing these droids; the mannequins provided material for experimentation. Note that these are white; in postproduction, George changed them to beige.) After short location shoots in Italy and Tunisia, we returned to the studio and wrapped on Friday, September 26. Pickups took place at Leavesden in September and November 1997, in March and August 1998, and for one day in March 1999.

During principal photography, I have a fairly low-key approach (right, with George on location). I only flag things for him that are really worth talking about; a lot of little potential problems I'll let go and deal with later, or I'll talk to a department head on the side. You also have to know that when George asks you to take two guys and shoot some second-unit scene in the parking lot, he's serious. He doesn't want to find out the following week that it turned into a twenty-person crew with a lot of expensive equipment. He doesn't like things to escalate. If he thinks the shot can be done with two guys in half an hour, he wants other people to think along those lines and follow through with that idea.

*The Phantom Menace* (1999)

WRITER-DIRECTOR: George Lucas
PRODUCER: Rick McCallum
PRODUCTION DESIGNER: Gavin Bocquet
DIRECTOR OF PHOTOGRAPHY: David Tattersall
DESIGN DIRECTOR: Doug Chiang
VISUAL-EFFECTS SUPERVISORS: John Knoll, Dennis Muren, Scott Squires
SUPERVISING ART DIRECTOR: Peter Russell
SET DECORATOR: Peter Walpole
EDITORS: Paul Martin Smith, Ben Burtt

## Republic Cruiser Cockpit

The Republic Cruiser was supposed to look a little familiar, as though it was made by the
same manufacturer as the Rebel blockade runner in *A New Hope*. The control panels (only a
few of the buttons are labeled—how does the crew know which ones to press?) were therefore
retrodesigned (opposite), as was the cockpit (below left and middle; below right: seen from
the outside, it's evident how small a set this was). In the cruiser cockpit is actor Silas Carson
(middle and top right—Carson is on left; actress Bronagh Gallagher sits next to him aboard
the *Radiant VII*), who also plays Neimoidians Nute Gunray and Lott Dodd, as well as Jedi
Master Ki-Adi-Mundi. This is the only time, however, that Carson is seen in *Star Wars* without a rubber mask over his face.

Although one of the first scenes in the film, the Republic Cruiser cockpit footage was
actually filmed on day fifty-one of the shoot, Monday, September 8, 1997.

## 120

## Trade Federation Ship Hangar

The Federation battleship hangar was built twice: once as a miniature about four-feet high, ten-feet across, and twelve-feet deep (opposite, with Brian Gernand and a model of the parked *Radiant VII*; below left, final frame). It was created in two sections, and this became the basis for the digital model that we see Anakin flying through during the film's climactic space battle. The later CG model looked great thanks largely to the fact that we'd built the miniature first, photographed it, and projected those images onto the digital reproduction. The second build was actually within the motion-control model of the Trade Federation ship (right, with Grant Imahara); we used this model for the shot where Anakin's starfighter goes out of control and spins into the hangar (top left, final frame).

## 121

## Trade Federation Ship Hallways

A number of hallways are depicted inside the Federation battleship. In standard *Star Wars* fashion, a very limited section of the set was built—just enough for the actors to do their scenes (below left and below right). The same set was then used for the hangar entrance, the hall outside the conference room (below left), and the hall outside the bridge.

Extensions of this environment were achieved with a ⅛-scale miniature. We built three sections of the corridor which, by adjustment of how they fitted together, were used to depict both straight and curved hallways. These were made to look longer than just three sections by the placement of a mirror (opposite; note the photographer in the reflection at the far end). The one-mirror trick was enough for curved halls, because the other end could be hidden behind the bend (top right, final frame). For straight halls, mirrors were placed at both ends and the camera pointed through a three-inch hole, which gave the illusion of any desired length we needed (middle right, final frame).

## 122

## Trade Federation Ship Bridge 1

The bridge was a fairly minimal set. After George decided that all of the live action would take place on the platform (right, discussing the bridge maquette with department heads), the extent of the set build was limited to just that area with three chairs (below right, on set in London). A ⅛-scale miniature was built in postproduction, which we could use in two parts—front and back of bridge—and was populated with CG battle droid pilots and a couple of bluescreen Neimoidians (opposite, the front of the model; top left, front of the model in a final frame; below left, the rear of the bridge model in a final frame).

# Trade Federation Ship Bridge 2

The bridge had to be filled with ship-status display screens. The question was: What kind of visuals would be of use to the droids that were "manning" many of the control stations (opposite, nonanimated display screens on set)? We decided that the displays should be skewed toward "machine readability"—i.e., barcodes and information designed for easy machine interpretation, but not necessarily intelligible to humans—therefore Noel Ruben in the ILM art department headed up the creation of dozens of droid-friendly graphics (below and right; below left, final animation on monitors).

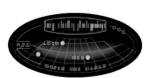

## 124

## Invasion Troops

After escaping the destroyer-droid assault outside of the bridge, the Jedi make their way through an air shaft back to the hangar where they arrived—there they discover an invasion fleet ready to attack Naboo (top right, a model of the H-shape troop carrier, with Danny Wagner; below left, final frame). They jump from an air shaft down to the floor below and conceal themselves among the ships.

The air shaft was originally going to be only about six feet up from the hangar floor (opposite), but George later changed the distance to about twenty-five feet. The Jedi jump was then filmed during the March 1998 pickups, and the actors were put on a wire rig. I worked out the physics of how many frames it would take for the jump, which in actuality was only twelve feet; we slowed them down just enough so their robes would flow nicely and give me enough hang time for the final movement. Most of the wide shots, however, used the hangar miniature (below right, final frame with raised air shaft at center of image), which was populated with hundreds of battle droids, troop carriers, and tanks.

## 125

## Theed Palace

The helicopter style-establishing shot of the Theed palace is typical of the techniques used in 3-D matte paintings. First, the scene was laid out in three dimensions, with very simple shapes to determine the composition. Once the composition and camera move were approved, more detailed 3-D models were built, but no more detailed than necessary. The final palace geometry model (opposite, top left and right) was textured with a relatively generic limestone pattern. Additional details were painted on top of the render in 2-D, and then projected back onto the 3-D model. Because the camera was flying in and to the left, most of the 3-D parallax is seen in the palace and the foreground cliff. The level of detail becomes progressively less toward the rear until, in the far distance, simple flat cards (yellow wire frames) were used (opposite, top left—these wire frames can be seen on the left of the image). A fairly traditional flat painting was done at this point, and projected onto the simple geometry. Photographs of miniature buildings, constructed for the Theed central plaza sequences, were used to fill in the background city (see spread 133). Waterfall elements were created by pouring salt over some velvet covered obstacles, and then photographed with a high-speed camera (right). When all these elements were combined and rendered from the moving camera, the painting now moved in very convincing 3-D (opposite, below: final frame).

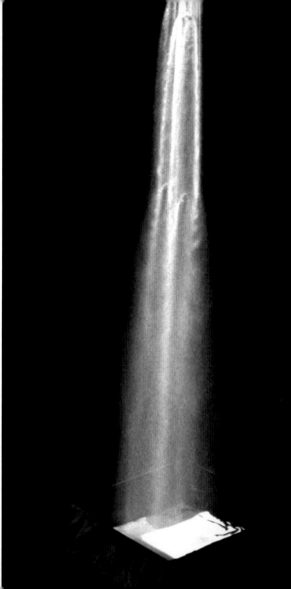

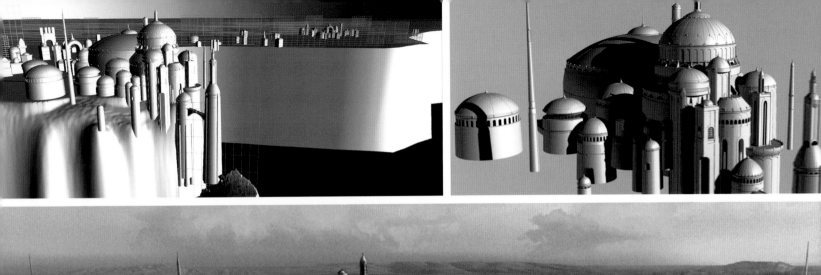

## 126

## Droid Landing Zone

The Trade Federation H-shape ship sets down on Naboo in a clearing adjacent to a woods and a swamp, and from its belly emerge MTTs. There was no location shoot for this scene. The backgrounds were all assembled in the matte-painting department from still photos of Whippendell Woods (which had been used for location shooting) and the hills of Livermore (located in the San Francisco Bay Area). Two different scale models were used for the MTTs: a model of the full ship with about an eight-foot wingspan (see spread 124), and a three-foot-high miniature of the front cargo bay (opposite, the model on a table; right, a close-up; below left, final frame).

## 127

## Naboo Swamp Stampede

The Naboo swamp was little more than some peat moss, water, and a few plants spread out on a plastic sheet in front of a bluescreen draped over one wall of a soundstage (opposite, Qui-Gon running; below, Lucas and crew setting up the shot). Later scenes in the Naboo woods were shot in Whippendell Woods just outside of London, and reference stills from those were used as the basis of the matte paintings created to fill in the background.

Back at ILM, Dennis Muren learned about a couple of trees that were to be cut down by Pacific Gas and Electric because they posed a danger to some local Marin County power lines. PG&E agreed to let us film these trees being pulled down, and these are the ones we see falling behind Qui-Gon Jinn (Liam Neeson) as he flees the invading droid army (right, final frame).

## 128

## Naboo Lake

Naboo lake is a combination of footage shot in the Whippendell Woods, the Pinewood Studios paddock tank, and the Action Underwater tank. There are no bodies of water within Whippendell Woods, but even if there were, we wouldn't have been able to film in one because of unknown parasites or other hazards that might be found in natural waters. For these reasons, shots of the Jedi wading into the water were shot in the Pinewood tank where water quality was known (opposite, a section of the Naboo forest was constructed within the tank, one corner of which is submerged—this is where the Jedi wade in; top right, final frame). The Pinewood tank was only about four-feet deep, however, so shots of the Jedi swimming down to Otoh Gunga were filmed at the Action Underwater tank in front of a greenscreen (below left, on the ESAB industrial estate), a locale about forty minutes from Leavesden. There the water was not exactly transparent—it was closer to translucent—so the scene appears slightly greener than we'd originally hoped (below right).

## Otoh Gunga Main Plaza

The Otoh Gunga main plaza set (opposite) was originally going to be used for two different scenes: one in which Jar Jar Binks (Ahmed Best) arrives with the Jedi and is confronted by Captain Tarpals (below left: Lucas prepares to shoot Tarpals's scene with a large foam-core "stand-in" for his CG steed), and another scene in which Jar Jar attempts to rally the Gungans after his return to Naboo, but finds his compatriots gone and the city ransacked. Although we shot the second scene, the footage wound up being cut. George felt it was sufficient to have Jar Jar come up out of the water and tell the Jedi no one was there.

The plaza set began its 3-D life as a maquette (below right) and was completed with a miniature and matte paintings (right, final frames).

■ 00270844

## Otoh Gunga Boardroom

The chamber where Qui-Gon Jinn and Obi-Wan Kenobi (Ewan McGregor) meet Boss Nass (below left, Brian Blessed performing his lines on set) consisted of a sculptured deck, two carved foam support braces, and a staircase leading up to the entrance (opposite). This set was built on Leavesden's MS1 Stage, and filming took place on Friday, July 18—the day before production left for location shooting. A ⅛-scale miniature was built for the rest of the chamber (below right, with Pat Turner), and combined with footage of the principals in postproduction (top right, final frame). The more distant Gunga city domes were computer generated.

## Through the Planet's Core

The Jedi's underwater submarine voyage was created by using the time-honored method of "dry for wet, foam rocks in smoke"—that is, the foam rocks are not actually placed in water, but special lighting and atmospheric effects were used so that when filmed, they appear to be (opposite). Gunga city and water particles were added with computer graphics, while the actors were shot in a half-cockpit submarine on an inner-tube rig at Leavesden (top right; below: final frame with "wet" rocks in background). As with the *Millennium Falcon* cockpit, the inner tubes allowed the set to be tilted and shaken by the crew.

## Theed Estuary

The Jedi voyage ends in Theed city's estuary. As filmed, Qui-Gon and Obi-Wan surface a few hundred feet from a waterfall and are unable to get their submarine to start (opposite). They struggle to get ashore before the ship is then carried over the waterfall. A full-sized submarine prop was rigged to travel along an underwater track in the Pinewood Studios paddock tank (top right, with crew on a dock in the middle of the tank). When the edit of the film was running long, however, this scene was dramatically shortened, leaving only a couple of shots from the sequence remaining in the film (below left, final frame).

In the background, Theed is a rearrangement of the miniature buildings that were created for the Theed plaza scenes (below right, model; top left, final frame).

## 133

## Theed Plaza

Theed city plaza was the largest exterior set built for *The Phantom Menace*, and was constructed on the studio backlot (opposite, preparing for a scene). It had to be big because a substantial number of shots were going to be framed entirely against this set, and for the most part they were without extensions (below right, a maquette—the white parts were built; the gray parts were designated as CG). Shots of the Naboo taken prisoner by CG battle droids (below left, on set) were completed back at Industrial Light & Magic with all the necessary CG and miniature elements (top right, final frame).

## 134

## Theed City 1

A handful of very wide shots of Theed city were extended with an elaborate miniature (below). About forty buildings between two and four feet tall were built for Theed plaza, so in order to fill up the backgrounds of other shots, they were continuously rearranged depending on the angle required (opposite page, the exterior of Theed hangar, with many of the plaza buildings repurposed; this view was also used for the shot in which an N-1 starfighter is shot down by a Trade Federation tank).

## 135

## Theed City 2

The Theed city miniature was shot outdoors on ILM's "backlot," otherwise known as "the slab" (opposite; below, on the set).

## Theed Hangar

Theed hangar was the largest interior set built for *The Phantom Menace* (top right, a production maquette shows how much was to be constructed). Still, the section on A Stage represents only a quarter of the floor area and a quarter of the height of the full hangar design as seen in the film (opposite, a gun battle between the Naboo and battle droid stand-ins; below right, the empty set). Fortunately, about sixty percent of the shots in this environment could be framed entirely against the constructed set and required no extension.

Extremely wide shots posed a big challenge, since the perspective was from a point outside the stage wall. Normally, the live-action portion of shots like these are photographed by a camera with a very wide lens placed as far back as possible. The plate is then shrunk down, turned over to ILM, and we're expected to "fill in the rest." The problem is that making an even wider shot out of something filmed with a wide lens yields unacceptable distortion. Usually a more-normal-focal-length lens is used instead, but then it's almost impossible to match perspective with the live-action plate. Ultimately, we completed these hangar shots by rotoscoping the figures out of the plate, replacing more of the set with a synthetic version, and then rotoscoping the actors back in (below left, final frame of the climactic duel).

## Royal Naboo Cruiser Cockpit

The cockpit of the Queen's starship was a beautifully designed, largely self-contained set (opposite; below left, cast and crew filming). It was also, from a visual-effects viewpoint, alarmingly replete with display screens—dozens and dozens of them! Everywhere the camera looked, at least five or six screens in varying sizes were visible. At our request, the art department, supervised by production designer Gavin Bocquet, made backlit transparencies with reasonably good-looking graphics on them. While the display screens weren't animated, when seen out of focus in the background, they looked pretty good (right, the backlit screens are below the three cockpit windows). This left us with the more manageable task of adding animated graphics to only those screens that were clearly visible in a shot, reducing our display-graphics workload by about seventy-five percent.

# Droid Hold

The royal Naboo cruiser's droid hold is where we first meet R2-D2 (right and below). Roger Christian shot most of this sequence with the second unit. The main unit filmed only the portion in which Obi-Wan locks Jar Jar in the droid hold "closet." Originally there was more of this sequence: The cruiser is hit by Neimoidian fire, loses a stabilizer, and starts spinning—with Jar Jar ending up stuck to the ceiling. The final cut has Jar Jar entering. Not much later, the astromech droids are called upon to fix the ship's shield generator. To do so, they take a droid elevator (right, center of image) to the ship's hull (opposite, on top right of image, a droid can be seen emerging from the set below via the "elevator," suspended by wires, to voyage out onto the hull; that sequence was filmed on a different set—see spread 139).

## Royal Naboo Cruiser Hull

In order to repair the shield generator and hyperdrive damaged by Neimoidian fire, a crew
of astromech droids, including R2-D2, travel out onto the hull of the ship to make repairs
(opposite, on set with R2-D2; the droid on the right is being supported by a back-brace
cable). For these shots, a section of curved hull, covered in Mylar film, and a four-by-eight-foot
section of exposed machinery was built. This set was photographed in front of a greenscreen
(right, the greenscreen is reflected in the Mylar) so we could composite it with photography of
the ten-foot-long ship miniature (below, final frame).

## Royal Naboo Cruiser Interior

The Queen's ship interiors were complete environments (below right and opposite) and required only that animated graphics be inserted for the display screens. You might ask why we didn't prepare these graphics in advance and then play them back on the monitors. The answer is that the irregular shapes of most of the screens meant that rear projection would have been the only practical way to get an image onto them—and some of the sets had thirty or more screens (opposite, facing chairs, are a few of them). The expense of the projectors and the systems required to motor them would have been high. It was far cheaper to insert the graphics in postproduction, where George could then decide exactly what animation was needed based on his edit (below left, final frames).

## Storm in Tunisia 1

We had been shooting in Tunisia for two days when a powerful storm hit during the night of Tuesday, July 29, 1997. The evening before, as we were all having dinner at the hotel, dark ominous clouds began rolling in. A little bit later, it started raining. As rain poured down harder and harder, lightning flashed all about and the wind grew stronger. As I lay in bed trying to sleep, I kept thinking about the sets. What was happening out there?

The storm was over by morning, but when we arrived at the shooting location, we were appalled at the devastation. The neatly laid out Podrace starting grid looked like it was the victim of an angry child who had tossed the Podracers into the air and then smashed them into the ground (opposite and below). Mos Espa was in pretty bad shape, too. Every set had been damaged to differing degrees (top right, I'm probably looking a little too happy in front of what's left of Mos Espa).

## 142

## Storm in Tunisia 2

Makeup and wardrobe tents were blown over, and Liam Neeson's Qui-Gon wig was half buried in wet sand. My first reaction while touring the damaged sets was that we would have to cancel shooting for at least a few days to give the art department enough time for repairs. Instead, Rick McCallum went into *über*-producer mode, got on his walkie-talkie, shifted the schedule around, and we moved to the least damaged set: the ramp to the Queen's ship. After about three hours of cleanup, we were ready to shoot. Meanwhile, the art department worked on the next-least-damaged set, and so on. In the end, by using this incremental method, we stayed on schedule and didn't lose even a single day of shooting.

## 143

## Tatooine Landing Site

The landing site was the set least damaged by the storm, and therefore required only minor cleanup for use on July 30. It really consisted of just a ramp and the ship's front landing gear (opposite, crew creating a path for R2-D2 by putting down sheets of plywood and covering them with a thin veneer of sand; anything deeper than about ⅜ of an inch, and the mechanical droid would get stuck; below left, final frame), both of which had been sent from Leavesden after the Coruscant landing platform scene had wrapped on July 3.

A ten-foot-long miniature of the queen's royal starship was used to complete these scenes back at ILM (top right). Because the ship was all reflections, and in order for its mirrorlike surface to be true to Tatooine's desert environment, we built a sandy tabletop surface below it (below right), and waited for the horizon to be clear of clouds before filming the miniature on the roof of ILM's main stage.

## Mos Espa 1

A sizeable portion of *The Phantom Menace* takes place in the city of Mos Espa on Tatooine. Therefore, multiple environments were needed: Watto's junk shop, front and back; Sebulba's café; Jira's fruit stand; and a number of streets and alleys. After George and production designer Gavin Bocquet discussed the size of the build (below, from left: Bernie Bellew, with arms crossed, director of photography David Tattersall, Bocquet, and Lucas), it was pretty clear that a couple of partial streets, with the key locations built into them, would cover the majority of the shots. One of these has the main characters entering the city, which was filmed on location (opposite, with crew); the raw plate (top right) was then combined with a miniature (middle right) to obtain a final frame (below right).

## 145

## Mos Espa 2

Mos Espa was one big T-shaped street created in Tunisia's Chott el Gharsa (right, photos taken of the main street set on location without enhancements). Although each location—the café, fruit stand, and junk shop—was pretty much right next door to the other, clever editing, set dressings, and camera work made them seem neighborhoods apart. About two dozen ¹⁄₄₈-scale building miniatures were built for the extensions (opposite, with ILM crew). Very wide shots combined numerous photographs of these buildings to create an entire city, and matte paintings were used for the distant backgrounds (below, final frames with enhancements).

## 146

## Watto's Junk Shop

The exterior of Watto's junk shop was built in Tunisia. The interior was filmed at Leavesden Studios in England (top right, Lucas on the undressed set; opposite, the dressed set)—once again editing joins two locations that were really thousands of miles apart. For views inside the shop looking out, a painted backing and bright lighting created the illusion of a bustling street. The lights also simulated the heat of the desert (below right, shooting out the doorway; below left, the junk shop interior set seen from the exterior on the soundstage). In order to create sightlines for the actors and for future CG reference, on-set photography made use of a foam-core cutout of Watto (below middle).

There was a big industrial air conditioner within the stage—but it couldn't be switched on because it was too noisy; the shooting also moved at a very brisk pace, so there were few opportunities to turn it on between setups. As a result, temperatures on this set reached almost 105° F.

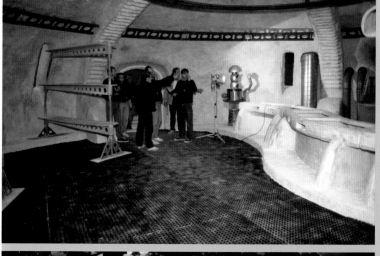

## Watto's Junkyard

The backyard of Watto's shop is where he keeps large pieces of salvageable hardware. The set was supposed to be complete in camera, but George decided, while editing the scene, that he wanted it to look bigger. In about half the shots, a wall of crates (top right, with Neeson as Qui-Gon and a stand-in with hat for the CG Watto; the hat was discarded in Episode I, but Watto is wearing it in his one scene in Episode II) was therefore removed to give us a wider view of the junkyard (below right, with CG Watto; also, compare the still, top left, shot on set, with the enhanced junkyard in the final frame, below left). These extensions were not done as matte paintings but instead created as touched-up photo collages. The images came from a variety of sources: the Tucson aircraft boneyard, the Lucasfilm stills archive, and photos of our Podracer miniatures. A pod reminiscent of the one in *2001: A Space Odyssey* (opposite) also appears.

## 148

## Slave Quarters

We shot all of the Tunisia locations from July 28 to August 10, 1997—when it was very, very hot. On one of our last two days in August, while shooting the scenes at the slave-quarters exteriors, I took a picture of a thermometer on an equipment cart in the shade (top left, with perspiring author top right). It read 53.7 degrees Centigrade, or 128.6 degrees Fahrenheit. In fact, before we'd left for Tunisia, production circulated memos to the crew warning us to stay out of the sun, wear long pants, long-sleeved shirts, sun hats, and at least SPF 30 sunblock. I took this seriously for the first couple of days, but it was just so hot I couldn't stand it. I switched to shorts, light cotton short-sleeved shirts, and no hat, and was much more comfortable. George, however, almost always wore dark plaid long-sleeved shirts and blue jeans. Amazing. I don't know how he survived.

The Mos Espa slave quarters was a dressed location in Medenine, Tunisia (below right and opposite). The building was originally a grain storage warehouse, and is now a hotel. The art department added *Star Wars*-style doors, control panels, and vaporators to create the Tatooine look. Back in 1976, this site had actually been considered as a location for *A New Hope*. George had always liked the looks of the environment, and decided to use it twenty years later.

## 149

## Anakin's Hovel

Like the interior of Watto's shop, the interior of Anakin's hovel was built as a set on Leavesden's D Stage (below right, being constructed). A brightly lit painted backing (top right) created the illusion that the apartment was adjacent to the Medenine location (below left, final frame).

Shmi Skywalker's (Pernilla August) workstation consists of her chair (she's sitting here when Anakin comes home after winning the Podrace), a microscope, and bits and pieces of hardware (opposite). It's possible she refurbishes high-end electronics for Watto.

## Coruscant Secret Balcony 1

Darth Sidious's secret conference with his Sith apprentice, Darth Maul, takes place in a seemingly unfamiliar part of Coruscant (opposite, on the balcony, the actors are filmed against bluescreen). To avoid building an entirely new set for what is essentially a very short sequence in the film, we shot on the balcony of Palpatine's apartment (see spread 151). By blacking out the windows and creating a background matte painting with unusual architecture, the balcony becomes a foreign environment (below left, final frame with Ian McDiarmid as Sidious and Ray Park as Maul).

## Coruscant Secret Balcony 2

The balcony of Palpatine's apartment (opposite; below, final frame) also served as the Sith's secret meeting place. This scene was shot on Thursday, June 26, 1997, the first day of principal photography, on Leavesden's Stage MS2.

## Slave Quarters Backyard 1

The slave quarters backyard, where Anakin completes his Podracer, was a partial build created near the Mos Espa set in Tunisia.

## Slave Quarters Backyard 2

Though the extent of the backyard set was large enough so that the majority of the sequence did not require extensions (opposite, on location with Natalie Portman as Padmé and Jake Lloyd as Anakin Skywalker; C-3PO is manipulated by his puppeteer Mike Lynch), a couple of wide shots were completed using photographs of the Medenine slave-hovel location as well as a matte painting (top right, final frame).

A nighttime scene (middle right) was extended as a miniature coupled with a matte painting by Bill George. The miniature's buildings were a series of modular pieces that could be assembled in a variety of ways to make an almost unlimited number of unique structures (bottom strip: three examples of rearrangements).

## Darth Maul's Landing Site

Darth Maul's Tatooine landing site background was a cliff edge overlooking a big butte known locally as "camel rock," near Tozeur, Tunisia (below right; sharp-eyed viewers might recognize the formation as the same location previously used in *The English Patient*, 1996). No set was built here at all. The one shot where Maul walks down the ramp of his ship was filmed bluescreen at Leavesden. His speeder bike was computer generated, so only a simple blue shape was provided for the actor to climb onto (opposite; below left, final frame).

## Podrace Hangar

This large hangar is where the racing teams prepare their pods for the Boonta Eve Classic. The set needed to be large enough for eighteen racing teams and their pit crews, which meant an environment far too large to build in its entirety, though a good chunk of it was constructed on Leavesden's B Stage; it included a number of full-sized pods—such as Sebulba's and Gasgano's—a sixty-foot back wall, sand, maintenance equipment, etc.

    The conversation between Qui-Gon and a computer-generated Watto (voiced by Andrew Secombe) takes place about halfway down the length of the hangar (opposite: top left, on-set plate; top right, photography of the background using a miniature; below, final frame completed with CG characters). To create the illusion of a complete hangar, and to cover the close-ups of Qui-Gon and Watto during their dialogue, the two directions behind each of the characters was shot using the same miniature—because, even at 1/8 scale, it was also very large, so only half its length had been built. In both cases, we're looking down the entire length of the model, which for each direction has been substantially redressed (below left, dressed miniature behind Watto; below right, dressed miniature behind Qui-Gon). Bluescreen elements and computer-generated characters populate the background (final frames, top right, and middle right).

## Podrace Arena

The Mos Espa arena was conceived as a huge Podracing venue capable of seating 500,000 spectators. About fifteen minutes of screen time, consisting of about 150 separate shots, takes place in this location. In the planning stages, it was clear that the arena was just too big to build, so only a few fragments were constructed as full-sized set pieces. This meant that any miniatures would have to work from multiple views and distances. A model of the arena was built as large as we could afford to; a working size of forty feet across turned out to be the limit (below right, with Bob Hill). Yet even that large a miniature, at about 1/72 scale, was too small for the tight shots. Therefore a handful of larger sections were built as "hero" miniatures: Jabba's box (1/4 scale), the announcer's booth (1/4 scale), the start/finish gate (1/4 scale), Qui-Gon's viewing platform (1/8 scale), and pit row (1/8 scale).

Computer-generated crowds were created for staircases, walkways, dignitary boxes, and the race track. The CG crowds were controlled with a few simple behavioral rules. They were programmed to move toward attractors ("carrots") and away from repulsors ("skunks"), and not to bunch up or collide with other "people" or architecture (below left and middle, CG people). In the establishing shot, carrots were placed at the stadium entrances and skunks were positioned around the Podrace starting grid (opposite, final frame; note that near the top of the final frame, a Jawa sandcrawler is looking for a place to park).

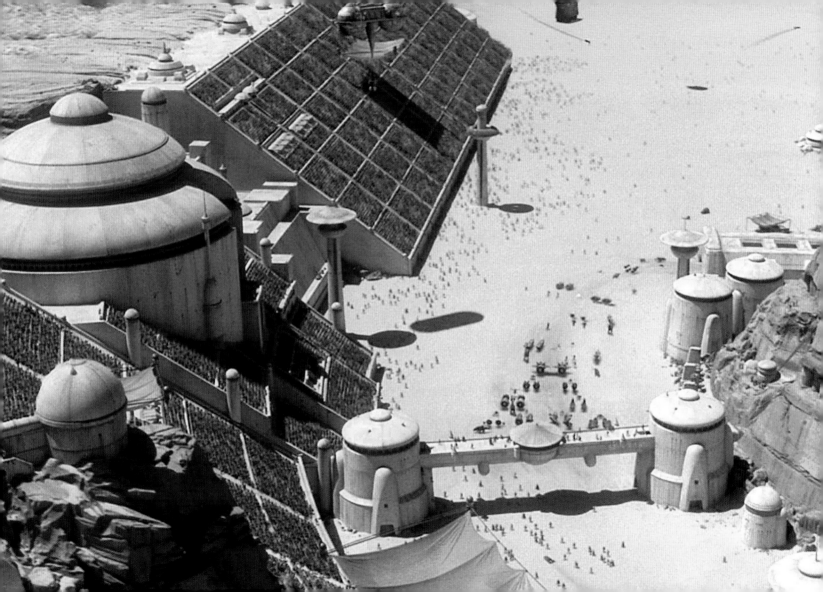

## Podrace Arena Stands

In Tunisia, one section of the arena bleachers was built full-scale for close-ups of extras (opposite). Of course, that didn't help us much in the creation of the hundreds of thousands of spectators. Consequently, we developed the following strategy: Medium shots of the arena would be populated by bluescreen elements of ILM crew members filmed on consumer DV cameras. The elements were recorded as individual rows, then layered into the stands to create a teeming crowd (middle right, final frame), or composited into plates shot in Tunisia (top right, final frame, with ILM staff filling out the crowd and ascending and descending the added stairs). Wide shots, where the size of an individual person would be extremely small, were accomplished with painted Q-tips, whose shapes made for the right kind of texture and were actually pretty convincing from a distance (below left, close-up of Q-tips; below middle, Mike Lynch and the thousands of Q-tips placed in the miniature; below right, final frame with Q-tips in stands).

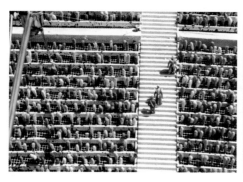

## Jabba's Private Box

Jabba's private box at the Mos Espa arena was not built as a full-size set; because Jabba was to be featured, but was a CG element, we created the box sequence entirely at ILM. First, we built a ⅛ scale model (below right and opposite, with Mike Lynch). Additional characters were shot as bluescreen elements, and there are a number of cameos in this sequence (below left, final frame with, from left: coordinator Bill Tlusty, in mask; George's daughter Amanda is the red-haired character just inside the arch; MatchMover Jack Haye; visual-effects producer Jeff Olson; animation director Rob Coleman; and supervising sound editor Matt Wood plays Bib Fortuna).

## Watto's Private Box

Because Watto has some important dialogue at the Podrace with Qui-Gon Jinn—a flesh-and-blood character—his box was built as an actual set (right). Watto's guests are all human actors in makeup appliances, or animatronic creatures, but he was computer generated (opposite: on set, Watto's physical invitees, including Warwick Davis as Weazel, are present but he's yet to be added into the shot; the large-headed creature was actually designated "Mace Windu" at the conceptual stage, before Samuel L. Jackson was cast in the part). Watto's box was constructed before Jabba's box or the announcer's booth miniatures, so it became a guide for their set dressing, textures, and finishes (below, final frame of Qui-Gon and Watto in the latter's box after the guests have cleared out).

# Announcer's Booth

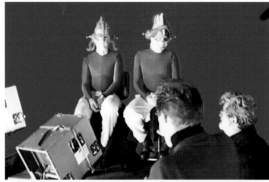

Since no first-unit actors appear in or around this location—and because the two-headed announcer was to be a CG creature (top left, final frame)—this booth was a natural choice for a miniature, which was built at ¼ scale for tight shots (below right, with Grant Imahara and Don Bies). Bleachers on either side (opposite; announcer's booth is center; below left, final frame) were populated by ILM crew, shot separately as bluescreen elements. Originally, the announcer was going to be played by two fairly well-known improvisational comedians in prosthetic makeup and with a CG body; we went as far as shooting them against bluescreen in our May pickups before it was decided to go with a totally CG version instead (top right). It was the simpler choice, because it did away with the blending of CG with live-action and its attendant details such as supertight matchmoves, balancing the lighting, etc.

█ 00235566

## 161

## Podrace Starting Grid 1

The starting grid consisted of nine full-size racing pods that were built in England and which had been used first at Leavesden for the Podrace hangar scene on July 8, 1997 (see spread 155). They were then flown to Tunisia aboard a giant Russian cargo plane, along with a handful of additional set pieces also constructed at Leavesden. The Podracers were set up in various formations for each scene, and filmed out in the desert of Chott el Gharsa on August 5 and 6 (opposite; below, group photo—cast and crew).

## 162

## Podrace Starting Grid 2

The second unit, headed by Roger Christian, shown here getting footage of the ground crew uncoupling the engines and running off to the side, as the Podracers roar to life. Note that some of the background pods are flat, painted cutouts.

## Podrace Starting Grid 3

The Podracers were supposed to appear as though they were hovering, à la Luke's landspeeder, and were thus supported by pipe frameworks during principal photography (below right, on location). In postproduction, we painted them out and added a few other elements to complete the scene (below left, final frame). The Podracing stadium was a combination of a matte painting with a ¹/₇₂ scale miniature (opposite, with Marty Rosenberg). The model was built so that the camera could be placed below the ground in a variety of positions in order to get the correct human-eye perspective at this scale (top right, final frames).

## Pit Lane 1

On location in Tunisia, Wald (Warwick Davis), Kitster (Dhruv Chanchani), and R2-D2 watch the big race from the pit lane.

## Pit Lane 2

When we shot the pit lane scenes on set in Tunisia, C-3PO was not in them (see spread 164). Originally, the protocol droid was going to appear only in Anakin's bedroom—but George kept adding him to scenes in postproduction, including this one. So we filmed C-3PO against blue- and greenscreen, with Mike Lynch operating the puppet, and composited him later into the picture (top left, final frame; Anthony Daniels later supplied the voice).

    Additional scenes looking across the starting grid into the pits, and a botched pit stop by Ody Mandrell's crew—a droid manages to get himself sucked through one of the Podracer's massive engines—were done using a ⅛-scale model (right and opposite, with author and a miniature of a Podracer; below left, final frame).

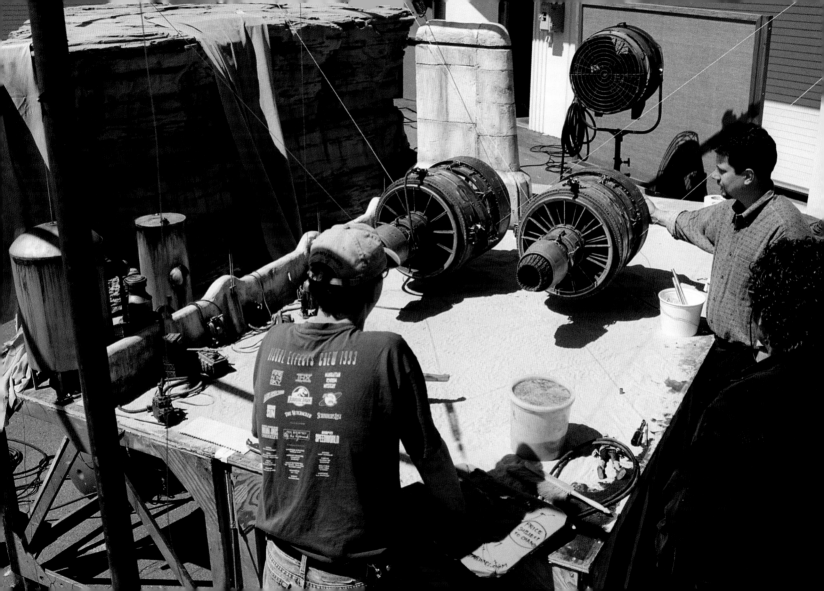

## 166

## Observation Platform

Qui-Gon, Padmé, and Jar Jar watch the race from an elevated circular platform in the arena.
Although a set was constructed in Tunisia (below left and opposite, with additional spectators
on the platform's left; the principals are on the platform's right), a miniature was used for the
shot of Qui-Gon's platform rising up from ground level (right: final frames of the miniature
platform's ascension combined with principal photography). Most of this sequence was
combined with footage of the stadium miniature in postproduction (top right, as seen in a
final frame).

## 167

## Podrace Course

A great deal of energy and painstaking work went into the design of the Podrace course. It was important that the terrain be a little exotic, with enough variety in the race's segments so that when the film cuts back and forth between racers, the audience knows where one racer is relative to another. For example, if we see Sebulba pass through Mushroom Mesa and out into Ebe Crater, and then a shot later we see Anakin entering Mushroom Mesa, we need to know visually how far behind he is. Robert Barnes therefore drew up a map of the conceptual Podracing course (opposite: the course is outlined in blue—it starts in the upper right-hand corner at the arena; proceeding counterclockwise to the Waldo Grade, which was an elevation change; next is Mushroom Mesa, which is followed by Rick's Rise, an elevation drop; a sharp left turn brings them to Ebe Crater, a big flat area with sinkholes that leads to the narrow slot through which is the curvy Beggar's Canyon; this leads to the Dune Sea portion of the race, but that was deleted; next is Arch Canyon, which leads to Jag Crag, where all the stalagmites and stalactites are, which goes into Lagulla Caves— where Rats Tyrell blows up; next is Tusken Turn, where the Raiders snipe at the Podracers. The last stretch is Hutt Flats, where Sebulba crashes. The lines in the middle were originally going to be an ancient abandoned city, with bits of old stone heads—those eventually were transposed to Naboo's Gungan sacred place). Creating distinctive terrain for the Podrace was therefore one of the biggest challenges in the film.

    Using a helicopter to shoot aerial plates of a desert location was considered, but much of the action required (such as weaving in and out of rocks and passing through narrow canyons), ruled out that approach. Miniatures were considered and were used for some of the more enclosed environments, but they weren't practical for the wider views because of how large they would have to be. In the end, that left only computer graphics. Coincidentally, Paul Huston and I had been experimenting with a 3-D matte painting technique that we'd used on *Mission: Impossible* and on the *Special Edition* of *A New Hope*. This involved projecting highly detailed textures (either photographed or painted) onto lower-resolution geometry, without lighting or shading. The 3-D rendering served to warp the images with the correct perspective as the camera moves. This technique could be freely mixed with more traditionally shaded computer graphics, making it extremely flexible. I thought we should try this approach here, so Paul put together some amazing test shots that convinced everyone this was the way to go (right—I was on set when I received the 8mm tape that Paul sent over of the test footage, and it was a big hit there; George said, "Wow! This looks great!").

## Mushroom Mesa

Mushroom Mesa was so named for the big mushroom-shape rocks that populate this area. Miniatures were built of a handful of these rocks (opposite, with a modeler sculpting one of them), which we photographed outdoors in our parking lot; the images were then projected onto simple geometry. This method provided the detail and realism of the model in a lightweight form that was easy to render.

Managing complexity was one of the big challenges in rendering the CG terrain of the Podrace. We needed detail in the ground near the camera—but that level of information would consume massive amounts of computer memory and rendering time if it also existed far from the camera. The solution was to have only the detailed geometry near the camera. A terrain-tool program was therefore written that looked at what the camera could see on a per-frame basis, and then created patterns only where the camera was focused. In the wireframe top view (far right), the red lines show the camera's field of vision. Note that high-density patches are near the camera, low-density patches farther away, and no geometry at all is created in the extreme distance (below left, the camera's view); the next steps were a preview render that's not motion-blurred (below right), and eventually the final frame with Anakin's Podracer in the foreground (right).

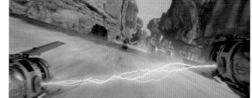

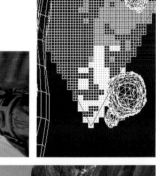

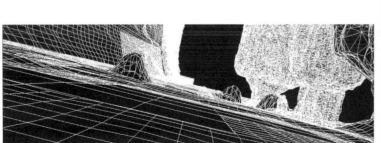

## Beggar's Canyon 1

Because of its high walls and extreme turns, we decided to build Beggar's Canyon as a
miniature (opposite and right; below left, final frame). Mixing miniatures and computer
graphics was desirable for several reasons: First, it helped distribute the workload. The
computer graphics department was so heavily burdened with other assignments, anything
that could be achieved using miniatures was a given. Also, if you have a wide open
sequence with no live-action plates, it's just as easy to build a miniature as it is to create a
virtual CG set; plus, a miniature is a good way to ground the location in "reality." I'm also
a big proponent of mixing techniques. If you create a sequence using only one approach,
the audience catches on and has time to spot the flaws. But if you vary the methods—
CG, miniature, CG, live-action, etc.—the audience is less likely to detect the visual effects
and, hopefully, will be more caught up in the action and the story.

## Beggar's Canyon 2

Preparations are made at ILM to shoot the Podracers' entry into the canyon's extremely narrow entrance (opposite; below, final frame).

# Arch Canyon

Of all the Podrace terrains, Arch Canyon is my favorite. It's exotic, but looks real. The racing action as the pods careen through it is incredibly fast and you can really feel the speed and danger. This was also the first environment for which we worked out many of the 3-D matte painting techniques that would be used throughout the sequence. About eight miniature arches were built, mostly about eighteen inches across. They were created out of foam and plaster, and had very nice, realistic paint finishes (below right, Steve Walton painting one of the models). Paul Huston carefully photographed these models in the ILM parking lot from a variety of angles and lighting directions (opposite and top right). He then used a device called a Microscribe digitizer to make low-resolution polygonal computer models that were reasonably faithful to the shape of the miniature. Projecting these photographs onto the CG models resulted in a collection of extremely realistic but lightweight models for rendering. Everything in the scene looked natural because most of what we were seeing was photography of real objects (below left, final frame).

## Jag Crag

The Jag Crag environment was built as a miniature for most of the same reasons that Beggar's Canyon was—high walls and extreme turns. It was also a closed-in curved environment with a limited view (below right), while the complexity of the stalagmites and stalactites would have created an impractical computer model. To shoot the miniature, the camera was hung down, cantilevered into the set from above through a crack in the ceiling (opposite, with Pat Sweeney and Mike Olague). An amusing detail on the model was a crashed Podracer situated on the side of the cave near the entrance; dust was added and the miniature was aged—the idea was this accident occurred a while ago and the wreckage was just left there (below left).

# Coruscant Landing Platform

When Padmé's royal cruiser arrives on Coruscant, it lands on a large floating platform. This site was far too large to build as a set, since it wouldn't have fit inside even the largest soundstage available to us. But because all the live action was to take place in the area between the landing ramp and the taxi dock, just a small section of the platform was actually constructed (opposite). For shots of the parked ship, we used a ten-foot-long miniature of the cruiser. The environment around the platform was created by matte painter Yusei Uesugi, who combined 3-D building models with 3-D matte painting techniques (right, final frames).

## Palpatine's Apartment

The scenes in Senator Palpatine's apartment were filmed on the first two days of shooting, Thursday and Friday, June 26 and 27, 1997. The second day it was raining heavily outside, which made for a lot of noise on the steel roof of the stage—I can only imagine what a pain that was for the sound editors.

From a visual-effects standpoint, Palpatine's apartment was fairly straightforward, and part of it was complete in camera (below right). A large matte painting, wide enough to cover all the views visible through its many windows, was shared over the whole series of shots (opposite: set with bluescreen outside the windows; below left: final frame with the matte painting behind Queen Amidala). One complication, however, was that all of the tables in the room were supposed to appear as though they were hovering. During shooting, they had blue legs supporting them, which we were supposed to remove in postproduction (opposite). For once the blue color wasn't intended to help with the legs' removal; it was simply a strongly contrasting color that was intended to remind us to paint them out.

## Jedi Council Chamber

The Jedi Council Chamber set was built six feet above the stage floor because many of the characters featured were to be puppets, operated from below (below left, under construction). The chamber is situated high atop one of the Jedi Temple columns, which were constructed as a miniature at ILM (opposite, with John Goodson, Tory Belicci, and Kim Smith). The chamber's windows were somewhat modified in postproduction when George felt that they didn't extend low enough and that their central pillars were superfluous (below right, with central pillars; top right, a final frame without). The Council would subsequently appear in Episodes II and III but, as usual, the set was too large to save, so only the chairs were stored for reuse.

## Galactic Senate Chamber

The Senate chamber, a huge cone-shape space, consisted of a central podium for the Chancellor (opposite, on set) surrounded by hundreds of floating pods docked in a pattern along the wall. The podium, a small section of wall, and two Senate pods (one of which was detachable) were built full scale (below right). The remainder of the chamber was extended with computer graphics.

The trickiest part of this environment was populating each of the pods with an assortment of exotic-looking delegations. Because there were hundreds of these seen from a variety of perspectives, we needed a large library of people at various angles to choose from. Ultimately the elements were filmed with an array of consumer DV cameras, each viewing the extras from multiple elevations. The extras were also positioned on a turntable and recorded at a number of rotation angles. To complete a shot, first we'd generate a CG render of the empty Senate pods (top right); a computer program, written by Steve Molin, then automatically selected the correct element from the image library, based on the angle of the camera in relation to the pod (middle right), choosing one of hundreds of cards (below left, an example of an element, with Ellen Pasternack and *60 Minutes*' Rome Hartman), to create a final frame (below middle).

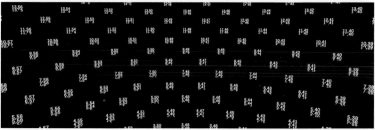

## Gungan Sacred Place

The Gungan sacred place, an old temple ruin, was a combination of plates shot at the Whippendell Woods location and a miniature filmed at ILM (below right, with Dave Fogler, with hat, Mark Siegel, and Dennis Muren). All of the actors and extras were recorded at Whippendell on August 21 and 22, 1997 (opposite), and a computer-generated Boss Nass was added to the composite. While the Whippendell location was beautiful, it wasn't particularly exotic, so an attempt was made to inject a little *Star Wars* into the miniature by adding large stone heads, a painted backing, and giant tree-root structures (below left, head in foreground).

## 178

## Gunga Grass Battle

The Gunga-versus-droid battle started as a shoot near Livermore, California. Several days of live-action plates and thousands of still photos were filmed. The matte-painting department modified the hills throughout the scene, stretching them vertically, to make them subtly taller and less Earthlike (top right). Hundreds of computer-generated Gungans and battle droids were then added in postproduction (top left). Layers of smoke, explosions (below right), and a computer-animated Gungan shield completed the sequence (below left, the force-field effect; opposite, final frame).

## 179

## Theed Generator Room

The generator room, though already a large set, was extended with computer graphics (top right, a live-action bluescreen plate; middle right, a CG render of the environment; below right, the final frame). Immense spaces with the same mechanical shapes repeated throughout are well suited for CG techniques since it costs nothing to make additional copies of an object. A miniature can be duplicated by molding and casting, but the costs are substantial.

The generator room was meant to be adjacent to the hangar (see spread 136), but that set already filled up a whole stage—so the door became a transitional piece (opposite, with bluescreen door; below left, a final frame with the hangar seen through the entranceway).

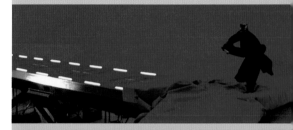

## 180

## Theed Palace Hallways

Typical for *Star Wars*, the palace hallways were all shot on one set (middle right, live-action shot), then made to look like multiple areas. In a number of cases, hallways were widened to make the space appear larger (top right, same set as middle right, but widened to seem like a different area). In addition, computer graphics were used to extend the columns and floor. Matte paintings were created for the far ends of the hallways (below right), and a miniature was built for reverse coverage of the droids (below left, with Marty Rosenberg).

Scenes with battle droids were also rehearsed with white-suited stand-ins (opposite, on set at the foot of the stairway). This silly-looking spectacle is useful for several purposes: First, it helps the actors know exactly where to look. Technically, the simplest way to record a scene that's going to have a CG character is to shoot it without anyone there—then there's nothing to remove or alter later. But problems surface when several live-action actors are supposed to be looking at that same CG character. If there's no place-holder stand-in, the live-action actors all have to look into empty space, imagining where the CG character might be—and often they end up looking in different places, which means the eye-lines are all wrong. Rehearsing with a stand-in helps alleviate this problem, because afterward, for the takes without stand-ins, the actors have a better idea where to look. The camera operator might also have difficulty composing the shot for characters who aren't present, and rehearsals with stand-ins are helpful here, too. Last, it's difficult to edit a scene without footage of all the characters. If we shoot a take with the stand-ins, the editor has a performance to cut with. For all of these reasons, scenes with CG characters are often shot both with and without stand-ins.

## Droid Control Ship

The film's climactic space battle centers around the droid control ship. A squadron of Naboo
N-1 starfighters attacks with the objective of disabling the droid army that controls Naboo by
destroying the ship. Most of this sequence's shots featured a six-foot-diameter miniature of the
entire ship (opposite, with Pat Sweeney), but this wasn't big enough for all of the plates. Two
more detailed sections of the droid control ship—the antenna array and the bridge—were
created for close-ups that skimmed along the craft's surface (below, final frames).

## 182

# Energy Beam Corridor

The energy beam corridor was composed of two sets: one with moving doors (opposite) and a shallow pit of about six feet; an additional set consisted of a deeper pit of about eighteen feet (see spread 183). The main set featured a mirrored floor. This surface posed some shooting problems because we had to be careful about what was being reflected there (for example, myself, top right). It also had to be kept very clean. In fact, during the filming of this scene, everyone wore white "booties" at all times—but it was still necessary to constantly sweep the floor free of dust (below right). This set also required additional work back at ILM, since any effects added—extensions, lightsabers, energy beams, etc.—needed to be created a second time for the object's reflection (below left and middle: final frames).

## 183

## Energy Beam Pit

The climactic duel finishes on the rim of a seemingly bottomless pit—which in fact existed as two sets of different depths: one of about six feet and another of about eighteen feet (right, under construction). The shallower one was used for shots such as Darth Maul falling after being struck down (opposite). The deeper one was called on for shots such as Obi-Wan's point of view, staring up at Darth Maul, or the duel seen from above (top left, final frame; below left, on set).

# Droid Control-Ship Hangar

The control-ship hangar presented a challenge in that we needed to see down a great length of it from a moving camera. Because we'd already built a smaller version of this space for the opening scene, we decided to make the hangar the best of both worlds: the realism of the miniature and the CG ability to duplicate it off into the distance. The earlier model was thus thoroughly photographed, and those photographs were projected onto simplified geometry in the computer. This gave us something that looked as real as the miniature (opposite, the final CG render), but which could be endlessly replicated, and which had no limitations related to camera access or clearance (i.e., set walls or ceilings or floors). Within a virtual set you gain valuable space because the virtual camera casts no shadow, and can be placed anywhere the director wants it (below, final frames).

## Droid Control Ship Reactor

When Anakin skids to a stop at the far end of the hangar in his N-1 starfighter, a second miniature was created for that back wall (below left, final frame). When he fires his starfighter guns at several approaching battle droids, one of the stray shots flies down a corridor and destroys the control ship's main reactor. We decided to build the reactors as breakaway miniatures. One highly detailed model was built, and a mold taken. The rest of the reactors were then cast from this mold out of a urethane plastic that had nice shattering properties (opposite, as one of the reactors explodes, others can be seen behind it). In an effort to economize, we reused the miniature of the hangar's back wall as the reactor room wall (right).

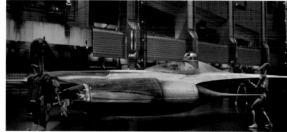

## Theed Palace Tower

The discussion between Obi-Wan Kenobi and Yoda (voiced by Frank Oz) about Anakin's future was added after principal photography had wrapped and editing began. George felt that the film needed this new scene, so it was added to the March 1998 pickups shot at Leavesden. The set was relatively small and built in its entirety. The only extension was the matte painting seen through the windows (opposite, without matte painting; below right, final frame with the painting in place).

    The tower floor was built five feet up from the stage and had a removable section in the middle to accommodate Frank Oz as he manipulated the Yoda puppet (below left). Yoda paces back and forth during this sequence, and I was amused to see that Frank wrote his dialogue along the inside of the hole in the floor, each line written where Yoda stops to speak them (below middle). It's also worth mentioning that the first-ever CG Yoda was created for this scene's initial wide shot (top right, final frame).

## 187

## Victory Parade

The parade on Naboo that concludes *The Phantom Menace* is a combination of live-action (opposite, photographed on the partial Theed plaza set at the Leavesden airstrip; note that only the lower halves of the buildings were constructed), and entirely synthetic shots created at ILM. The Theed plaza miniatures were used to extend the backgrounds.

Maximum use of 200 or so extras was made by shooting them grouped in different parts of the real set and melding them together. Synthetic shots began as a plate of the miniature (top right). Extras consisting of ILM employees and their families were photographed as bluescreen elements, again shot in groups and replicated to form the crowd (right, second down). The Gungan parade was all computer graphics, with three levels of detail used, depending on their size in the frame (right, second from bottom, mini Gungans; bottom right, large-scale Gungans). The simplest models were positioned in the far distance, which saved us time rendering (below left, final frame).

## 188

STAR WARS: EPISODE II

# DIGITAL EVOLUTION

## Changes

The really big innovation for Episode II was digital cameras (opposite, one of the digital cameras, with "B" camera operator Calum McFarlane). We were abandoning film and going digital, so we talked about what the requirements might be. Being able to download a "response curve" was one thing ILM really wanted.

All digital cameras start with an image sensor. The sensor reads the voltage, and that voltage is turned into a number. Between the voltage and the number is a mapping—the "response curve" of the camera. What we wanted to do is load our own response curve into the digital cameras. At the time, ILM was using an 8-bit standard called 8-log-38, which was a proven way to encode into eight bits of data a reasonable representation of the visual portion of a film image. My scheme was to load an 8-log-38 table into the high-definition (HD) camera to match our system. But about five weeks before we were going to start shooting, after many revisions, we went with the Rec.709 response that was already loaded into the camera by Panavision. This response was originally meant for video; even though we had wanted a filmic response curve, it ended up working out reasonably well.

The digital cameras also created new elements on set: for example, the tented abode of high-definition (and quality assurance) supervisor Fred Meyers, whose "mother-ship" technical supervisor Michael Blanchard helped create (far right, with Blanchard, HD decks, and the computer that controls all the camera settings); and the video village, Lucas's mobile command center, now had hi-res, fifty-inch plasma monitors (right). It was just great to see these preview monitors on the set—they were linked directly to the digital cameras, so we knew pretty much exactly what we were recording.

The other big change was that production moved to Fox Studios in Sydney, Australia. Principal photography began there on Monday, June 26, 2000. On Wednesday, August 30, we began our location shoot in the Lake Como region of northern Italy, moving to Tunisia on Wednesday, September 6. After one day in Spain, we finished stage shooting at Elstree Studios in London, England, from Friday, September 15 to Wednesday, September 20—all in all, a sixty-one-day shoot. Pickups were filmed in London at Ealing Studios in March and November 2001, with an additional day in both January and February 2002 at Elstree Studios.

*Attack of the Clones* (2002)

DIRECTOR: George Lucas
PRODUCER: Rick McCallum
STORY: George Lucas
SCREENPLAY: George Lucas, Jonathan Hales
PRODUCTION DESIGNER: Gavin Bocquet
DIRECTOR OF PHOTOGRAPHY: David Tattersall
VISUAL EFFECTS SUPERVISORS: John Knoll, Pablo Helman, Ben Snow, Dennis Muren
CONCEPT DESIGN SUPERVISORS: Doug Chiang, Erik Tiemens, Ryan Church
SUPERVISING ART DIRECTOR: Peter Russell
SET DECORATOR: Peter Walpole
EDITOR: Ben Burtt
DIGITAL MATTE SEQUENCE SUPERVISORS: Yusei Uesugi, Paul Huston, Jonathan Harb

## 190

## Coruscant Landing Platform

*Attack of the Clones* begins when a starship and several N-1 starfighters arrive on the planet Coruscant. The entourage thinks they've arrived safely—when an assassin's bomb blows up the starcruiser. The Coruscant landing platform set was pretty sparse, just a big rectangle and a ramp, but to accommodate the story, the art department dressed it in several ways: with some mist and a couple of ladders for when Padmé (Natalie Portman), disguised as a pilot, descends from her starfighter (below, on set; top right, final frame); the moment of the explosion (bottom right, on set; second from bottom, final frame with digital explosion); and post-explosion with small debris and smoke (opposite, on set; right, second from top, final frame with set extension and added large debris). Neither the N-1 fighters nor the droids (except for R2-D2) were there on set; they were added in postproduction.

## 191

## Palpatine's Office 1

The office of the Supreme Chancellor had a really nice *Star Wars* feel, with a great wraparound wall (below right; below left, with his makeup and prosthetic on but without his costume, Matt Rowan as Orn Free Taa waits, appropriately, in Palpatine's waiting room for the rehearsal to begin). George once explained to me that throughout the six films, a red-to-gray progression occurs within the succession of Palpatine abodes. The first time we see his office, in Episode I, it's totally red. In Episode II, Palpatine's office has more gray in it: a gray desk, a gray floor. In Episode III, the office has even less red. Gradually the color is going away so that, eventually, the only color left will be embodied in the Red Guards. If you look at the Emperor's throne room in Episode VI, it is an all gray-and-black environment.

A couple of months before principal photography began, I had decided to buy a good digital still camera. I'd been waiting for years for a solid, reasonably priced camera that could shoot 2K pictures with interchangeable lenses and decent exposure settings. Fortunately, the Nikon Coolpix 990 came out—and coincidentally one of the accessories was a fish-eye lens. For *The Phantom Menace*, I had manually stitched together panoramas from photos taken with a wide lens. We'd needed these panoramas in order to create reflections in the eyes of digital creatures—so that Watto's eyes, for example, would reflect the room he was in—but it was a painful and distorted process. With Nikon's new accessory, I realized I could write a software program that would enable me to capture a whole environment with just a couple of digital pictures (opposite, for example, Palpatine's office). So two weeks before shooting in Sydney began, I wrote the fish-eye "unwrapper" and was eventually able to create hundreds of panoramic photos of the sets.

# 192

## Palpatine's Office 2

A panoramic point of view from behind Palpatine's desk (opposite; below, final frame).

## Padmé's Apartment Exterior

For shots of Padmé's apartment exterior, not only did we construct the building—we also created a mini interior, which included tiny scaled furniture and sculptures (opposite, with Adam Savage). The model had various uses: the first establishing shot (top right); the moment Obi-Wan Kenobi (Ewan McGregor) crashes through the window to grab an assassin droid (middle right); and another shot in which a droid replaces the window that the Jedi so heroically shattered (bottom right).

## Padmé's Apartment Interior

The interior of Padmé's apartment set was built twelve feet up from the stage floor to provide enough greenscreen coverage, given the floor-to-ceiling windows (opposite), and to allow Obi-Wan's jump through the bedroom window toward the droid. It's here that Padmé and Jar Jar Binks (Ahmed Best) are reacquainted with Obi-Wan and Anakin Skywalker (Hayden Christensen) at the movie's outset (below, final frames).

## 195

## Padmé's Bedroom, Coruscant

A couple of scenes take place in Senator Padmé Amidala's bedroom, which was a fairly complete set with only the windows and closet requiring extensions. In one sequence, an assassin droid introduces poisonous insectlike creatures into the room in order to complete the botched job on the landing platform (below left). Movie sets are often surreal: In one part of a large stage, you might have some fairly intimate drama being played out, or someone pretending to sleep, but if you look just slightly to one side you'll almost always discover a crew of about a hundred people doing their jobs (opposite: ever try to sleep with fifteen people watching you?).

Another sequence has the Senator preparing to leave Coruscant because of the multiple attempts on her life. The day of shooting that scene on the bedroom set, George decided that the door of the closet should be open and that Padmé should be getting the clothes for her journey (below right, on the set; top left, final frame with the extension). The suitcase she's packing, which we later see Anakin carrying on Naboo, is perhaps one of the least exotic looking props in the *Star Wars* saga. George did have second thoughts about the luggage, but it was too late in the schedule to modify it.

## 196

## Parking Garage

Anakin "borrows" a hot rod airspeeder to aid his Master in tracking down Padmé's would-be assassin. The other airspeeders were all computer graphics (opposite, on set; below, final frame).

## Zam's Ledge

Bounty hunter Zam Wesell (Leeanna Walsman, opposite, on a ledge without her helmet, while we set up for a take) is about to try and shoot Obi-Wan Kenobi out of the air as he clings to the assassin droid. The ledge she's standing on was a small piece of set with animated signs supposedly behind her. Because the signs were going to be used as a light source for the set, I asked production designer Gavin Bocquet to build something into the walls that would provide the right kind of interactive light. He had back-lit panels constructed; in postproduction, we added in the graphics, while the wider shots were completed in the matte painting department (below, final frames).

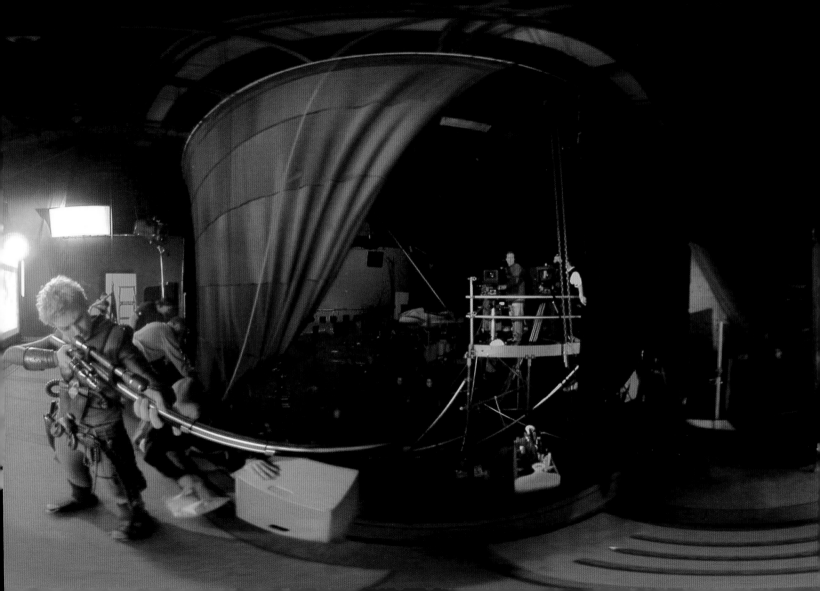

## Zam On the Go

Zam Wesell in the cockpit of her airspeeder, as she flees the Jedi (opposite, on set; below, final frame).

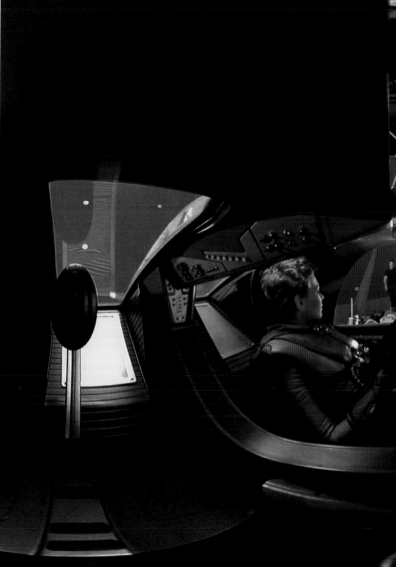

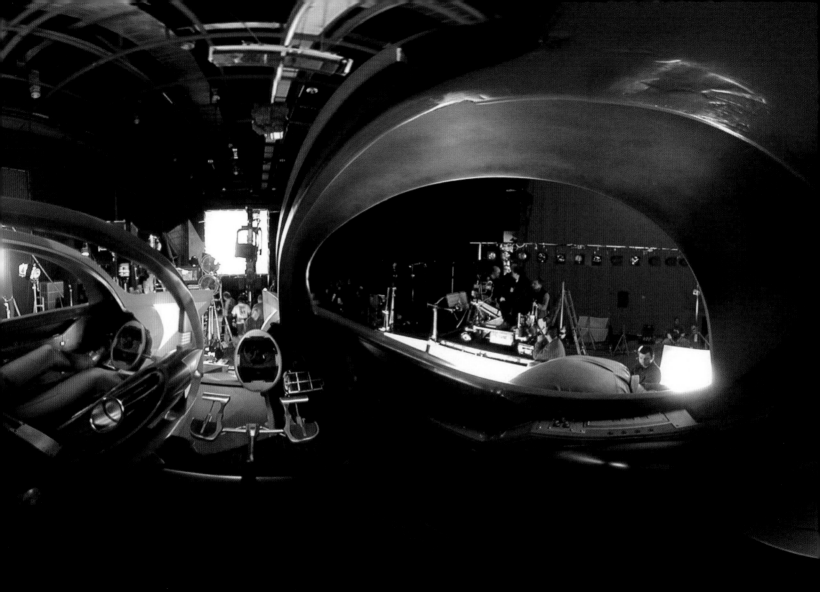

## Airspeeder Chase

Zam Wesell jumps in an airspeeder, as Obi-Wan is picked up by Anakin in another airspeeder, and the chase is on (second from top right, a final frame. If you look closely you'll see an in-joke: a creature known as a shaak is flying by on the lower right). The only set pieces during the entire pursuit through the Coruscant skylanes are these two airspeeders, both of which were shot on rigs (opposite, it's Hayden's turn on a bungee-cord rig; below, a panorama of the interior and instrument panel of his airspeeder from the driver's point of view; below right, Zam's airspeeder on a hydraulic-pipe rig; top right, a final frame with the camera positioned so that it looks like the actors are flying downward). We all assumed that Anakin's airspeeder—with its yellow paint job, exposed engine, and generally hot rod look—was a homage to the yellow sleek machine that Harrison Ford drives in George's *American Graffiti* (1973).

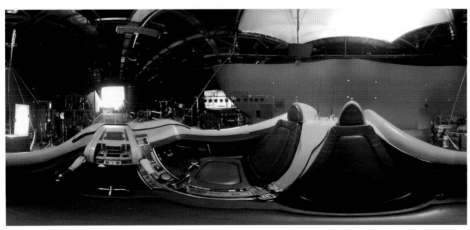

## Speeder Chase Environments

In the same way that George wanted the Podrace to traverse a variety of terrain, he asked for the airspeeder chase to go through different neighborhoods. He was very specific about this. His logic was that if you were watching a big car chase go through Manhattan, you wouldn't see the same generic buildings whizzing by; the architecture, built over hundreds of years, would change constantly. So, after a ten-minute chase through a real-life city, you'd be in a totally different place from where you started. I'd imagined that we would build a tool kit of generic-looking buildings that we'd rearrange as needed throughout the chase—but George wanted us to push the concept even further. Instead we built many, many assets for each specific neighborhood—and the airspeeder chase sequence took us a little over a year to complete.

We start the chase near Padmé's apartment, in an upscale neighborhood befitting the ruling class, which consists of nicer, newer penthouse-style apartments sporting rounded art deco spires (below left). Wesell then leads Anakin and Obi-Wan down into the older parts of Coruscant, which are dominated by square, more Gothic-looking buildings (below middle). Next they traverse a financial district (top right), with glass façade buildings, and an industrial area, replete with power generators (second from top). They speed through a warehouse district, with big storage areas and immense cranes and big trucks (third from top), finally crash-landing in the entertainment district, with all of its appropriately wild Las Vegas strip colors and animated signs (below right; opposite, Robert Barnes' reference map).

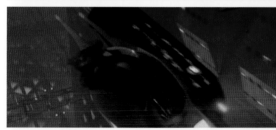

NOTE: ONLY BUILDINGS' HEIGHTS TO SCALE - OTHER ARCHITECTURAL
PROPORTIONS NOT ACCURATE FOR REFERENCE

## Speeder Chase Traffic

The first time we tackled Coruscant skylane traffic was at the end of the *Special Edition* of *Return of the Jedi*. The Coruscant celebration had to have airspeeders and other aerial circulation, so I had CG modeler Colie Wertz build a dozen shapes of really low-res vehicles, about 250 polygons apiece. It was very primitive. To avoid having some two thousand objects in my scene, we arranged the traffic in strings, so that it was all laid out and saved as one model. I couldn't have individual action, but by having one string move slightly faster than another, we achieved a little parallax motion.

On Episode I, we dealt with circulation in fairly similar ways, because we never had traffic moving particularly close to camera. It was always something way in the background. We got through all of Episode I doing it that way—but when we got to Episode II and we found out about the airspeeder chase, which was going to take place right next to a lot of flying vehicles, we knew we were going to need better models and more sophisticated traffic behavior. To do it effectively, I wanted each car to have its own speed and to have lanes just like on a freeway, with fast lanes and slow lanes and passing lanes. We also wanted to have lanes that were perpendicular to one another, with some vehicles peeling off and joining a different lane. Development lead Neil Herzinger took all that on for Episode II. He helped us create a whole library of cars, some new and some from Episode I, and he wrote a tool that enabled us to lay out a path in space—straight, curved, etcetera—and then populate it with cars (opposite). With Neil's program, we could set up parameters: minimum and maximum speed, and vehicle types. In the warehouse district, for example, we decreased the quantity of passenger cars and increased the number of trucklike vehicles. After we'd decided on the parameters, the computer randomly distributed the cars along the path and gave them motion.

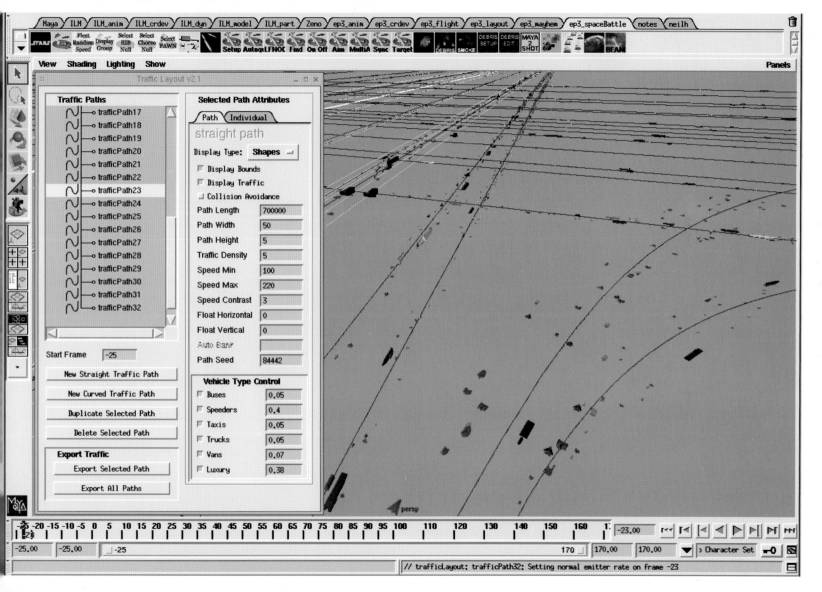

# Entertainment District

For the district where the airspeed chase ends up, we relied on a dozen miniature buildings that had removable pieces that could be moved around, extending what we shot in Sydney, Australia (below—in fact, this street is the redressed parking garage; right, final frames). By redressing these assets and lighting them differently as we changed angles, those buildings were multiplied into many dozen. But if you look closely you can see in the entertainment district things that aren't supposed to be there, like Dexter's Diner and a little archway containing Padmé's apartment (opposite, with model-makers Aaron Haye and Peggy Hraster).

## Nightclub Exterior

This set was built because of some important dialogue that Anakin and Obi-Wan have at its entrance. Our goal was to build just enough so that the scene could be shot entirely "against set," and to extend the wider shots with a miniature (opposite, with Peggy Hraster). Note the bluescreen through the entrance (below, panorama). The nightclub interior set was so large that it filled an entire stage—with no room for the exterior. Scheduling what sets (there were more than fifty) went on which stages (there were only six) to be filmed when (the shoot lasted sixty-one days) was a logistical jigsaw puzzle (other sets were occupying Fox Studios's larger stages on the days we shot the nightclub scenes). In the handful of shots where we see in through the entrance, matching angles were filmed of the interior set, and the exterior and interior were later composited together (right, final frame).

## Nightclub Interior 1

I always think of this as an example of the tragedy or pathos inherent in set building and dressing. These very elaborate creations are built over a period of months, and then destroyed as soon as production is done with them.

I saw the art and construction departments at work on the nightclub interior during a period of two months; I'd walk by Stage 3 day after day and see people lovingly crafting details and gradually painting bits and pieces of the carpentry. The day we arrived on set—Monday, July 17, 2000—I admired all the custom neon and Plexiglas tubes, the welded steel, and the incredible bar. It was a very detailed artwork that massive amounts of energy had gone into. It even had big 1960s-style pneumatic tubes (opposite, two of them can be seen; below right, a detail; during the first wide shot in the movie, if you look closely, you'll see someone who zips down one and steps out of the tube into the scene). We shot on this complex and intricate set for half a day. We got there in the morning, started shooting, and moved onto another stage by lunchtime. The next day, the crew tore apart the nightclub interior. (Top right, a detail of the drink dispenser.)

## Nightclub Interior 2

While we were shooting the nightclub scenes (opposite; top right, Obi-Wan Kenobi in conversation with Elan Sleazebaggano), George was saying, "I need to do something to this nightclub to give it a particular character—maybe a band? But we did that in the Cantina (in Episode IV). Maybe dancers? But we've done that, too (in Episode VI)." What he decided afterward in postproduction was to make it a sports bar or gambling club. Taking advantage of a moment when the Jedi come in and Obi-Wan looks around, George took that pause and cut into Obi-Wan's point of view—which became one of these gambling alcoves. George also decided that the betting room should have big-screen TVs, and asked us to visualize the kinds of sports they'd be watching: "You can do robot football, and you can do some kind of creature racing, and something else." I'd been playing LucasArts' video game *Racer*, where you can pilot Podracers on different planets, so I thought of putting in a Podrace as the third sport. We went though *The Empire Strikes Back* archives and found two plates we could use in some stock footage of a glacier in Norway originally shot from a helicopter (for the sequence in which the snow speeder is searching for Han Solo and Luke Skywalker). We matchmoved them and resurrected a couple of Podracers to complete the two shots.

The gambling alcove itself was built as a miniature, and I filmed ILM crew as bluescreen elements that were later composited in (bottom right, final frame). But I had a limited number of people—just five—so we had the same five people go through different wardrobe changes and become twenty people. In fact, visual effects producer Jill Brooks is in the same shot in three different costumes!

## Nightclub Alley

The walls used for this alley were multipurpose, as was the entire set. First it was painted with dark colors made dingy, and dressed with lots of garbage as the alley in which the Jedi interrogate bounty hunter Zam Wesell (opposite; below right, one of the dressings was an abandoned astromech droid), who is silenced by a poison dart fired by Jango Fett (top right, final frame, establishing shot with extensions; middle right, final frame with CG Jango). The alley then became the entertainment district street on which Anakin chases Zam: its walls were widened from sixteen to forty feet, it was repainted, and extras were recruited (see spread 202). Finally, the street became the garage where Anakin "borrows" the yellow airspeeder that he uses to chase Zam (see spread 196). It turned out that the sets were created in reverse order of the story.

# The Ultimate Council Build

Samuel L. Jackson (Mace Windu) awaits recording of the scene in which the Jedi report on the assassination attempts (opposite, on set; below, final frame).

## Jedi Analysis Room

Deep in the heart of the Jedi Temple is the equivalent of an FBI analysis room where various Jedi can do research and where Obi-Wan goes to investigate the origins of Jango Fett's poisonous dart (right, final frame). There is even a "clean room" on the other side of the window. This set was actually a redress of the Jedi Council chamber: the little wall dividers were pulled out, three consoles were put in, and the floor was repainted (opposite and below; note that all the reverses are matte paintings); the clean room is a miniature. So while we had to rebuild once again a locale that we'd used on a previous film, the Jedi Council was economically viable because ultimately we turned it into three additional sets: the analysis room, the younglings' training room (see spread 214), and Mace Windu's original office (see spread 229).

## Bus Terminal

Having been persuaded to flee covertly from the danger on Coruscant to the relative safety of Naboo, Senator Padmé heads for a space freighter dock aboard city transit, accompanied by her Jedi bodyguard Anakin Skywalker (top right, final frame). For the bus itself, we built half of its interior (opposite, bottom) and a full-sized exterior section (opposite, top), but not the cockpit. The interior's set dressing was up to the Sydney, Australia art department's usual level of detail (middle right, final frame), even including a Coruscant "subway map" (below). When the characters step out of the bus, the set is a miniature, with bluescreen extras composited in and a cityscape matte painting (bottom right, final frame).

## Dexter's Diner Exterior

When Obi-Wan Kenobi, as originally scripted, fails to find the origin of the dart at the Jedi Temple analysis room, he travels to a diner where he hopes to glean some information from his multi-worldly friend, Dexter Jettster. The sequence starts with a very nice matte painting by Yusei Uesugi (top right, final frame). The Sydney art department actually built a partial exterior of the diner in Sydney, Australia (opposite), and though we filmed it, we ended up using a miniature of the diner for the only shot from the outside (bottom right, final frame).

# 211

## Dexter's Diner Interior

This was an interesting set because, stylistically, Dexter's Diner is simultaneously a throwback to Mel's Diner in *American Graffiti*, while still containing futuristic elements (opposite, with bluescreen windows). One of them, the robot waitress, was an afterthought; originally a human waitress was going to be the only server in the restaurant. As for Dexter—a four-armed CG character—animation director Rob Coleman and I persuaded George and Rick McCallum that the more reference we could get on set the better. For the major CG characters, we really wanted to have an actor's performance to work from; George would have to direct this actor, which would be a big help later on. In Dexter's case, actor Ronald Falk brought a lot to the performance: a limp, character nuance, etcetera. The animators directly lifted those elements for Dexter's CG performance (below, final frames with Dexter and Coruscant life seen through the windows).

# 212

## Dexter's Kitchen

Although we see Dexter's kitchen only through its circular window, the art department built a complete set (opposite and below).

## Jedi Library

After Dexter informs Obi-Wan that the dart was made by clones who live on the remote planet Kamino, the Jedi asks librarian Madame Jocasta Nu (Althea McGrath) if she knows where that planet might be—since the Jedi Archives are silent on the matter (below left, final frames). All we built for this set was a piece of floor, a desk, and a couple of columns (below right); the rest was a miniature (opposite, with Pat Sweeney and key grip Tom Cloutier). The busts seen on the pedestals include a number of ILM cameo appearances, including visual effects supervisor Pablo Helman and animation director Rob Coleman.

## Youngling Training Room

This set is pretty obviously a redress of the Jedi Council chamber. The floor pattern was changed, and some columns and plants added (opposite). Only half the chamber was redecorated, which restricted the number of possible camera angles. The area outside the training room where Obi-Wan Kenobi appears is a miniature.

For Episode I, we'd populated the Jedi Temple with background characters in our usual way: by just putting some Jedi robes on ILM employees. But George complained about our uninspired method, saying all the Jedi looked like they came from Marin County (which they did, more or less). "They should be a much more diverse group," he explained. "You should have aliens in there, more exotic-looking people. So whenever you're doing the backgrounds in the future, you gotta mix it up more." So for Episode II, we made an effort to create more diversity.

Coincidentally, as we were preparing the extras for the training room shot, George mentioned a character who was really popular in the Dark Horse comic books, saying, "I think it'd be kind of amusing to see her in the background." So the Ranch sent us a bunch of the comic books featuring Jedi Master Aayla Secura, and we scoured ILM for someone who might incarnate her. Among our staff, production assistant Amy Allen looked the most similar, and she was game to do it. Next, Gillian Libbert created a costume based on the one in the comic book, and we proceeded to put Amy in the background (right, final frames). Before we knew it, George decided that she should be in the Geonosis arena as well, and that she should be one of the survivors—so her character just grew.

## Refugee Ship

On their way to Naboo, Padmé and Anakin get to know each other aboard a freighter ship carrying refugees. Things are getting bad in the Republic and many people are heading toward more isolated, safer regions of the galaxy. Again, the interior of the ship was a partial set, built in the art department depot of Fox Studios's Building 48 (opposite, in the foreground is the cafeteria-style layout, where R2-D2 grabs some food for the young couple; top left, final frame). The edibles were made from real materials, dyed and colored. The greenscreen (opposite, in background on the left) was originally going to be replaced by an animal pen filled with exotic CG creatures—*Star Wars* cattle—but that idea went away when the scene was shortened and the dialogue changed. The new lines were actually shot bluescreen during one of the pickups at Ealing Studios (below right, with Hayden Christensen and Natalie Portman) and combined with stills of the set (below left, final frame).

## Naboo Plaza

Arriving on Naboo, Padmé and Anakin disembark and walk to the Queen's palace. Their short trek was shot in Seville, Spain, on the Plaza d'Espana, which had been built for the Fair of Americas in 1929. On our way to London to finish stage work after filming in Italy and Tunisia, our one-day stop in Spain was on Wednesday, September 13, 2000, with a reduced unit: we didn't want to bring the whole crew for just a single day of work. The plaza was actually horseshoe shaped (bottom right, seen from above), but in the wide establishing shot we mirrored it to create an oval. The buildings were also modified in the comps: towers were altered with matte paintings, and domes were added (opposite, the actual plaza; top right, final frame).

The Plaza d'Espana was the only location we shot on that was publicly accessible, and word got out well before we arrived that George Lucas was going to be filming *Star Wars* there. So when we pulled up in our bus the morning of the shoot, there were literally three or four thousand people—screaming! Across the street from the plaza is a big public park, and the police had let fans come up to the park's perimeter fence. Each time someone got off the bus, a roar went up from the crowd. For a while they were chanting, "Where-is-Na-ta-lie? Where-is-Na-ta-lie?" By the time we'd wrapped, after our usual twelve hours, only a few hundred fans remained; George went and shook their hands, going all the way down the line. Ahmed Best did the same thing and so did Hayden Christensen, also signing autographs.

# Naboo Throne Room

Padmé and Anakin have a brief audience with the Queen of Naboo in the same palace we saw in *The Phantom Menace* (opposite, in the Caserta Palace in Italy, George, 1st AD James McTeigue, and 2nd 2nd AD Paul Sullivan discuss a shot). Once again, the art department had to do minimal set dressing: a bluescreen, desk, and chairs. We weren't allowed to take down the antique candelabras on the columns, so we painted those out of the shots later. Oliver Ford Davies (Gov. Sio Bibble, below left), because he's very much in demand as a theater actor, was unavailable while we were in Italy, so we recorded him later at Ealing Studios against bluescreen. (Below right, the monitors show us in split screen the scene we're trying to line up with; bottom, final frame.)

## Padmé's House, Exterior

Omitted from the picture, but seen on the Episode II DVD, is a scene that takes place in the house of Padmé's parents. George liked a particular painting a lot and wanted to replicate it in a shot where Padmé and Anakin enter her childhood home. The canvas depicts a street scene, with the staircase to the house on the left, an alleyway on the right, and children playing in the foreground; so our composition was based on this painting (bottom right, final frame).

We built the staircase and the floor in Sydney, Australia. I'm a proponent of creating as much as is reasonable of a set, because there are some things that are just really hard to replicate in CG. One of them is light reflections on actors' faces. Light bouncing off bluescreen onto the actors—what is often recorded on set—doesn't look effective in the finished film, particularly if behind the actors is a yellow CG wall. At first, production designer Gavin Bocquet was going to paint the wall and staircase blue. But in this particular case, Gavin told George that it wasn't any more difficult for his department to paint the steps and wall blue or another color, and that putting in the textures wouldn't be that hard or time consuming, so Rick McCallum and George gave it the okay. The result was a more satisfyingly realistic light for that scene (opposite).

Another scene that didn't make it into the movie, and which didn't make it onto the DVD either, was shot outside the Villa Balbaniello in Italy. It was supposed to take place outside Padmé's house, and had her father Ruwee Naberrie (Graeme Blundell) questioning his daughter's suitor, Anakin, about his intentions (top right, in Italy).

# 219

## Padmé's House, Interior

Constructed entirely on a soundstage, only the bluescreen for the interior dining room (opposite) and kitchen (below) can be seen through the windows of the dining room. We filmed the plates for these views at the Villa Balbaniello during our location shoot in Italy. This scene, in which Padmé's parents and siblings tease the Senator about her infatuation with Anakin, was cut.

■ 00168424

## Padmé's Bedroom, Naboo

Padmé's bedroom in the house of her parents was a real room at the Villa Balbaniello in Italy. A large-screen "TV" is covering up what was probably a painting above the fireplace (opposite). Hayden Christensen is on the bed; the period furniture was taken out, while the more Naboo-like rug and bed were imported (below, final frame). When the scene was included on the *Attack of the Clones* DVD, it was completed by Dan Gregoire and his animatics team at Skywalker Ranch.

## Kamino

Thanks to Dexter and the Jedi younglings, Obi-Wan learns that the poison dart was made by "cloners" and is able to locate the remote planet of Kamino. All of the Kamino sets were pretty minimal. Obi-Wan's landing platform was the biggest construct (opposite, the set; right, final frames with extensions and effects). The locale was also dressed to be the platform where Obi-Wan fights Jango Fett (Temeura Morrison). Because it's always raining on Kamino, sprinkler heads were installed, and the whole set was rigged for water: the construction department built a dam around the platform with eighteen-inch walls, so we wouldn't flood the stage. The art department had in fact chosen Stage 1 for the fight, because this particular soundstage had originally been built for livestock. (Fox Studios in Sydney, Australia, was once a fairground. The stage had been sloped for easy drainage because the locale was hosed down after each animal show.) Nevertheless, the rain on set was a bit miserable: the air became extremely humid, and we had to wear raincoats all day. Lots of extra effort had to go into keeping the camera clear and, by the end of the day, the camera crew and actors were drenched (below).

## Lama Su's Office

Upon arriving in Kamino, Obi-Wan is surprised to learn that he was expected, and is shown into the office of Lama Su, the prime minister (voiced by Anthony Phelan). We shot all of this environment bluescreen in Sydney, Australia, except for a blue shape for the chair. Unusually, this set and its particulars were honed in postproduction. George decided to back off from the concept art idea, where he'd imagined the office enveloped in a blinding bright light somewhat like the white-limbo prison in *THX 1138*. As with all of the Kamino interiors, the office was a mixture of miniature (opposite, with Pat Sweeney) and CG: the office and descending chair were models, while the adjoining hallway and sliding door were CG (right, final frames).

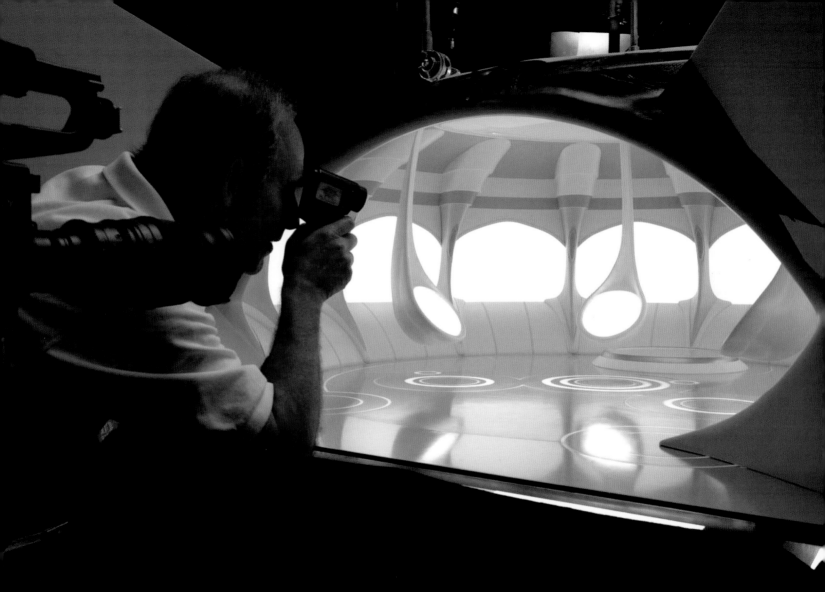

## Naboo Lake Retreat

The Villa Balbaniellon, an essentially undressed location, stood in for the Naboo lake retreat (opposite). It's here that Anakin and Padmé hide away from their assassins. A scene that was shot but later cut occurs after the couple's starship has landed on a platform on the far side of the lake. They were supposed to get into a gondola (top right, Anakin and Padmé; bottom right, the gondola) and speed across the lake, but in the movie we just see them arriving at the retreat.

# 224

## Villa Balbaniello

The villa, located on the shore of Lake Como in northern Italy, was the site of the location shoot for many scenes in *Attack of the Clones* (opposite; below, final frame).

## Kamino Clone Factory

The Kamino clone factory scenes were filmed completely against bluescreen with only a vague idea of what the shots would end up being. George showed Ewan McGregor a couple of concept-art paintings, and said to him, "Down over there, you see a lot of clones." So it was a major imagination exercise for the actor, the kind of thing I don't think he enjoys. The corridor where Ewan was walking was transformed thanks to matte paintings and two miniatures (opposite, with Marla Selhorn taking camera measurements of one hallway; top right, final frame; below left, Pat Sweeney working on the second corridor; bottom right, final frame).

## Naboo Picnic

Paul Huston did the matte paintings for this scene. Because he used to be a model-maker, he likes to build miniatures and then shoot them (opposite, Huston working on the miniature); the photos serve as the basis for his painting and he gets very good results that way. Waterfalls were one of the added elements; for those, Paul traveled to some natural falls he'd chosen and recorded them on video. They were then comped into the final scene (top right, final frame). The principals—Hayden Christensen and Natalie Portman—were shot on location in Italy. We found a big field on a hill with some nice grass a mile or two from the Villa Balbaniello, but the grass is probably the only thing that's left in the final frame from that shoot (bottom right, on location).

# 227

## Lake Retreat Interiors

During their seclusion, Anakin and Padmé share a dinner and afterward talk about their future. Recorded on July 19, 2000, before we left for Italy, both sets were built on Stage 4. The dining room had a bluescreen hung outside its windows (opposite, top; top right, final frame), while the next scene was nearly complete in camera (opposite, bottom; right middle, final frame). The eating scene featured some food supplied by our ever-resourceful art department (bottom right, note the striped fruit).

## Jango Fett's Apartment

Meanwhile on Kamino, Obi-Wan is introduced by Taun We to bounty hunter Jango Fett—the genetic original for all the clone troopers. His apartment was a fairly complete set (opposite, a view of the exterior), but you can't get through a *Star Wars* locale without some bluescreen. Here it's used behind the windows (below); later we added a gloomy sea, rain, and occasional lightning (right, final frame). In this case, the corridor really is right outside the apartment, which has a den and a little closet for Jango's armor; only one wall of the closet was built—just enough to accommodate Obi-Wan's point of view as he catches sight of the incriminating evidence. (Below, Rena Owen, who voiced Taun We, wears a photo cutout of her character attached to her plastic hard hat so that the actors had an eyeline when talking to the CG Kaminoan.)

# An Office and a Bed

An additional redress of the Jedi Council chamber was the original Mace Windu office. But this set was abandoned when we moved the scene to a more monklike meditation room (see spread 230). George didn't like the idea of Mace sitting at a big executive desk, because the Jedi seemed too terrestrial-based, too much part of a bureaucracy (opposite, with Samuel L. Jackson and a Yoda stand-in sitting across from him; below right, Lucas sitting behind the desk). On the other side of the same set, a four-poster bed was moved in for the scene where Anakin is awakened by a nightmare at the Naboo lake retreat. The shot was really close up, so no other set dressing was needed (top right).

## 230

## Jedi Quarters

After moving him from his office, Mace's quarters became much like, or nearly exactly like, Yoda's quarters: a hexagonal room with window slats, which we built 1/4 scale and used twice (opposite, with model-maker Tony Preciado). To differentiate Yoda's quarters from Mace's, we painted the former's yellow (bottom right, final frame), and the latter's a rusty red (top right, final frame). The only set dressing for each were the tuffet chairs (below, on the blue set).

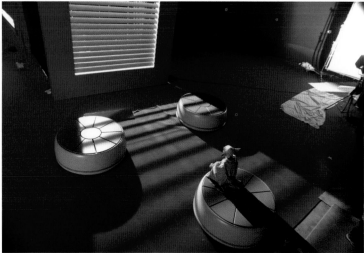

## 231

## Mos Espa Arrival

Anakin's nightmares about his mother, Shmi (Pernilla August), prompt his return to
Tatooine. He arrives with Padmé aboard her ship in an establishing shot, in which the
skiff turns and lands in a docking bay (below left, final frame). To fill in the background,
surrounding structures—low-rent, not terribly detailed—were salvaged from Episode I,
and combined with a matte painting and about a dozen new miniatures (opposite, with
model-maker Dave Murphy). Anakin and Padmé then take a rickshaw through the Mos
Espa streets to Watto's shop (below left, in the final frame the transport is visible just
outside the hangar). The day of the shoot, the vehicle was pulled by a pickup truck; later
on we painted out the rickshaw's wheels so it would appear to be hovering (below right,
rickshaw with wheels; top right, without wheels).

## Mos Espa

A panorama of the Mos Espa set. Sebulba's café is down the street on the left (opposite), Watto's shop is toward the center, and the archway forms the town's entrance (below, final frame).

## Watto's Shop

The main street of Mos Espa in *Attack of the Clones* is exactly the same one we built and filmed on during the location shoot for *The Phantom Menace*. Rick McCallum and George had made a deal with the Tunisian government, so we were able to leave it all just standing there—the result was that the Mos Espa set became a tourist attraction in the interim. People came from all over to see the *Star Wars* set!

Not only did tourists abound, but nomads had also made camp where the Naboo cruiser had been parked, about one thousand feet from the Mos Espa set (bottom right, the nomads; top right, the vessel's landing ramp shot on location during Episode I, in almost the same place as the nomads). When we returned after a three-year hiatus, some repairs had to be made. But after the second shoot, we left it standing again; essentially George donated Mos Espa to Tunisia.

For the scene outside Watto's shop, shot on Sunday, September 10, 2000, we recorded the principals under awnings (opposite). Director of photography David Tattersall didn't want the actors doing long dialogue scenes under the really harsh Tunisian sun, especially because it creates a lot of black shadows. On Episode I, a shot of Natalie Portman in the exposed desert had been somewhat upsetting for David: she was too brightly top-lit, and her eyes had been transformed into black sockets. At the time, we couldn't do anything about it; in postproduction, we fixed her eyes. But this time around David requested awnings so his lighting would work on location (below, final frame).

## Slave I Cockpit

The cockpit was the only part of Jango Fett's ship, *Slave I*, that the art department built (the ramp was just a blue rectangle). Originally it was designed so the set could be turned from a horizontal to a vertical position, because when *Slave I* is parked, the cockpit is actually perpendicular to the ground (below, final frame of cockpit interior). However, the day of the shoot, the art department said that it would need about an hour to rotate the cockpit from horizontal to vertical—and George didn't want to wait that long. He pointed out that Daniel Logan (young Boba Fett) could simply pretend that he had to climb into the pilot's chair. So that's what we did, and it turned out to be convincing.

Next to the cockpit is one of our big rotating light rigs (opposite, on right of image). That is, the light is rigged onto a car, which can be rolled on a semicircular track. Crew pull on a cable to slide the light around, which simulates moving sunlight and creates those moving shadows that George likes so much. In fact, throughout all six movies, whenever someone is in a flying starship, the shadows are moving—which makes it seem like the Y-wing, X-wing, TIE fighter, or *Millennium Falcon* are actually maneuvering in space.

## Jedi Starfighter 1

For Obi-Wan's aerial dogfight with Jango Fett, once again the only portion built of the Jedi starfighter was its cockpit (opposite and bottom right, on set). For other scenes, a full-sized starfighter was needed. Originally we shot Obi-Wan setting out for Kamino from within the lollipop-shaped Jedi hangar (below and top right), but that scene ended up being cut (though it was completed for the DVD). The cockpit, fairly complete in the full-sized ship, was used only for shots of Ewan McGregor getting into or out of the ship. For scenes in the ship's interior, it was easier to shoot and light a detached, fully dressed cockpit.

## Jedi Starfighter 2

A panorama from within the full-sized ship. Note the rock wall of the Geonosis landing site on the left, and the Tusken Raider camp on the right, all built on the same stage (opposite; below, final frame of *Slave I* cockpit).

## Geonosis Landing Site

After Obi-Wan eludes Boba Fett, he lands on Geonosis (opposite, with Ewan McGregor; note the visible Tusken Raider hut on the left). For that scene, we also used a full-sized starfighter. But we actually ended up shooting the full-sized model twice. The primary reason was because George changed the site of Obi-Wan's capture from atop a Geonosian spire—which we shot in Sydney, Australia—to just beside his ship, which necessitated a re-shoot of the starfighter for the first pickups (bottom right, the starfighter on a small area of dressed ground). But between shoots George also decided that Obi-Wan's ship should suffer a series of strafing holes during his dogfight with Fett. The second time around, his ship was therefore redressed to look more distressed (compare the cut final frame, top right, with the final frame used in the film, middle right, which features the damaged starfighter).

## 238

## Tatooine Homestead 1

The creation of this set was really fun. Visiting the original dry lake bed outside of Nefta in Tunisia—exactly where George had filmed Mark Hamill (Luke Skywalker) contemplating Tatooine's twin suns in *A New Hope*—was incredible. The crater that the earlier crew had excavated and shot on Monday, March 22, 1976, was still there on Thursday, September 7, 2000 (below). Here I was at this iconic place that I'd seen in a film that had changed my life while I was in high school. Of course twenty years of rain and sand storms and erosion meant that the Episode II art department had to reenforce the berm a little bit; the "igloo" had to be rebuilt as well (opposite). Thanks to advances in digital filmmaking, in Episode II the crater lip and homestead pit (which is actually hours of driving away from the lip) were seen together in one shot (right, an enhanced photo composites the two disparate locations into one place).

## Tatooine Homestead 2

A dawn panorama of the homestead on the Chott El Jerrid, Tozeur, Tunisia (opposite; below, Joel Edgerton as a young Owen Lars).

# Tatooine Homestead Center 1

Another really fun moment was revisiting the homestead center. After we wrapped at Nefta, production traveled in a convoy to Matmata. On paper, the drive looked like it wouldn't be more than three hours—the progress report even estimates it at 2.5 hours. But what looked like an hour on the map became in reality three hours, as we started slowly climbing through an incredibly twisty region. When we finally arrived at the Sidi Driss Hotel after about five hours of driving, on the evening of Sunday, September 10, 2000, the thing that really struck me was that the homestead center environment was mountainous and not at all what you see in the final film.

On the wall behind the front desk of the Sidi Driss Hotel is a floor plan, where visitors can see that there are several pits, each with dining rooms built into the walls (below, the map; opposite, the 1977 *Star Wars* pit dressed for *Attack of the Clones*).

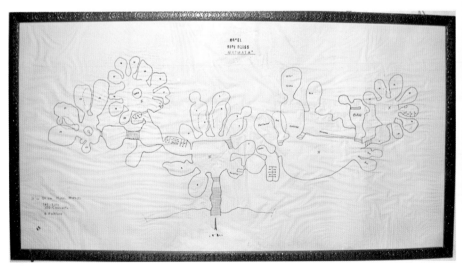

# 241

## Tatooine Homestead Center 2

Matmata, as it actually appears, with hills and modern buildings, and the homestead-center pit (opposite; below, final frame).

## Tatooine Homestead Center 3

A panorama of the center, with a stage for the real hotel's performers (opposite, on the right, behind ladder)—this stage isn't seen in any of the *Star Wars* films. On the left is the opening to the breakfast-nook dining area (below, final frame).

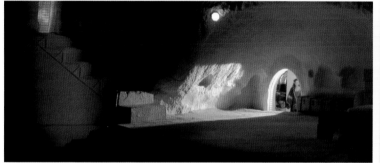

## 243

# Tatooine Homestead Kitchen

The Episode II homestead kitchen was a re-creation of the Episode IV kitchen (opposite; below, final frame). During set construction in Sydney, Australia, Gavin Bocquet's art department didn't have a lot of pictures for reference, mostly just frame blowups. No one had thought to save detailed photos because no one knew the first *Star Wars* was going to be so successful. Only one good photo was found, which the art department tacked up next to the site, so all the building crew could use it for reference while working on the set.

The other handicap to creating and shooting this set was its location in the art department building, not on a genuine soundstage. Fox Studios normally has seven stages, but while we were there the enormous Stage 7 was being used for the Titanic Experience—a tourist attraction using re-created sets from that film—and was therefore unavailable. The result was that our producer Rick McCallum had to scramble sometimes to find space for the sets. The homestead, the refugee ship interior, and a few other sets, were built in the art department depot—which had no soundproofing on the walls, and no lighting grid. Each time a plane went by overhead, we had problems!

## 244

## Tatooine Dining Area

I'd heard a story that a French fan had made the hotel Sidi Driss an offer, which was accepted, and had the alcove's ceiling decoration repainted to match the way it looked in *A New Hope* (opposite; below left, final frame). When I looked closely at the ceiling and walls, I discovered spiders living in what seemed like bored holes (below right). I was also able to verify that the dining area contains steps that lead to a storage closet, which in the films are supposed to lead to the kitchen.

When we first shot the dining room scene, C-3PO was still a puppet (opposite, on left). The idea was that Padmé, while waiting for Anakin to return from his search for his mother, has nothing to do so she puts on his coverings. That idea was discarded, however, and the protocol droid has his metal platings on throughout Episode II.

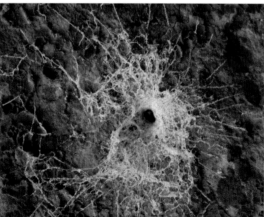

## Geonosis Spire

Although originally scripted for more scenes, the spire was ultimately used only for Obi-Wan's spying on a Separatist meeting. For the spire's interior, Ewan McGregor was filmed on some blue steps in London during pickups, and that plate was combined with a model (below right, final frame) and a matte painting (below left, final frame); for the exterior, a matte painting was composited over a photograph of a model because we needed to add more details (right, matte painting; opposite, the model, with Alex Jaeger, kneeling).

## Geonosis Conference Room

When I arrived in Sydney, Australia for the beginning of principal photography, I went with George on a set walkthrough. When we got to Palpatine's private conference room (bottom right, being constructed; top right, concept art reveals what it was going to look like), George turned to Gavin Bocquet and Rick McCallum, and said, "Umm . . . This scene is out of the movie." The set was already two-thirds built—and the whole conversation just stopped. But then George added, "I have another conference room scene now on Geonosis. If you make this one into a Geonosian conference room, we can still use it." The walls were already built, so they were ripped out and replaced with more Geonosian-style ones, and the floor was modified (opposite).

# 247

## Speeder Bike Set

On a soundstage with Hayden Christensen as Anakin Skywalker on a blue speeder bike (opposite, stunt coordinator Nick Gillard is on left; below, final frame).

■ 00136560

## Tusken Raider Encampment

Anakin finds out at the homestead that his mother has been kidnapped by Tusken Raiders. He borrows a speeder bike and sets out in search of her, coming to an encampment of the wandering nomads. The camp—two full-sized huts, dirt, sand, and a campfire—was built on a Fox Studios soundstage (below). In postproduction, we made it look larger by constructing a tabletop miniature with more huts (opposite, with ILM model-maker Pierre Maurer and chief model-maker Barbara Alfonso). It's here that Anakin takes his first decisive step toward the dark side, slaughtering the entire tribe in revenge for their murder of his mother (right, final frame combining the set in the foreground with the miniature in the background).

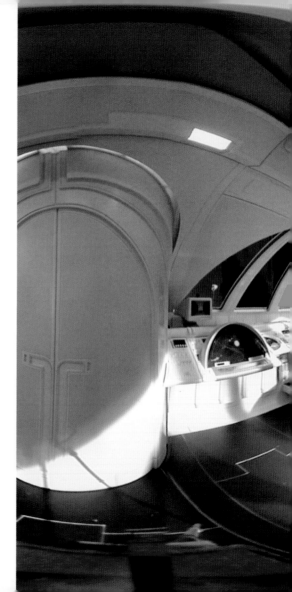

## Padmé's Starship Cockpit

While waiting aboard Padmé's skiff for Anakin's return, R2-D2 receives a holographic message from Obi-Wan—at the end of which, the Jedi is taken prisoner by destroyer droids. The starship was built with a lot of flyaway pieces so we could get the camera in where George wanted (below, seen from the exterior). After the principals had been recorded, sound designer/editor/second unit director Ben Burtt moved in to shoot the scene with R2-D2 as he plugs in (opposite, with senior model maker Don Bies sitting, droid controls in hand; Burtt is kneeling).

# Tatooine Homestead Center 4

This panorama is from a deleted scene in which Padmé wakes up during the night, walks over to the garage, and finds Anakin (opposite, Padmé is on left). It was actually shot during the day, with large black silk material spread over the opening to block out the light (below, final frame of Padmé in garage).

## Homestead Garage

Another fun re-creation of an Episode IV set was the homestead garage. For Episode II, this is where Padmé finds Anakin, who confesses to having slaughtered the Tusken Raiders. The original garage was bigger because it continued through the two doors, and the whole set was complete in camera; but George elected not to build that stuff the second time around (opposite). Therefore the views through these doors were completed with matte paintings and a CG model of a skyhopper (below, final frame). The hopper was designed with more rounded shapes, as if it were an earlier model of the same ship seen in Episode IV.

George's shooting script contained a scene in which R2-D2 plays the hologram of Obi-Wan's capture for Anakin in the garage (in the film, the scene takes place on the skiff ), echoing the moment he would later play a hologram for Anakin's son in *A New Hope*. The art department had therefore made a hologram-sized cutout of Obi-Wan (right) so Hayden would have an eyeline.

## Shmi's Funeral

One of the things we added to Nefta's dry lake bed was a mini cemetery, with three gravestones, about one hundred yards to the side of the homestead igloo. Of course the freshly planted one is for Shmi, Anakin's mother, on which the art department added some strange glyphs (opposite). The attendees included Shmi's husband Cliegg Lars (Jack Thompson) sitting on a chair (below), which we made to look like it was hovering in the final scene.

## Geonosis Prison Cell

The scene in which an imprisoned Obi-Wan speaks with Separatist leader Count Dooku (Christopher Lee) was added to the pickups at Ealing Studios (opposite). Essentially, ex-Jedi Dooku is telling Obi-Wan the whole truth about the Sith plot to take over the Republic, which is really cool because the way the story's been set up, Obi-Wan can't help but think Dooku is lying. On the set, Ewan McGregor was hanging on a flying rig in a standard two-point harness. In postproduction we painted out the wires and added the force field as a CG effect (below, final frame). Because the scene included a few pages of dialogue, and because Dooku circles around the set a few times while speaking to Obi-Wan, the crew had to make sure that the big high-definition cables didn't get tangled up (right).

## Geonosis Landscape

Some of the exterior terrain on Geonosis was accomplished with a miniature for various ground-level shots (opposite, model-maker Robbie Edwards works on the miniature; bottom and middle right, the miniature in two final frames). Other exterior terrain was completed with computer graphics, as in the shot where Padmé's ship approaches the cooling towers (top right, final frame).

# Geonosis Cooling Tower

As originally written, after Padmé and Anakin decide to go to Geonosis and rescue Obi-Wan, they basically walked through the front door and negotiated with Count Dooku, who arrested them because they wouldn't go over to his side. But George decided that this scenario was anticlimactic and that there needed to be something more dramatic. Therefore their confrontation with Dooku in the conference room was jettisoned and a series of action scenes were inserted in its place for the pickups. Now they arrive clandestinely, landing in a cooling tower (below left, final frame), and exiting onto its landing platform—one of our more minimal Ealing Studios sets, made up of hexagonal floor tiles and a doorway (opposite, on set; below right, final frame with the couple walking through the doorway).

## Sleeping Quarters

After exiting the cooling tower, Anakin and Padmé walk down a corridor known as "the sleeping quarters"—again at Ealing Studios, with just some more floor tiles and a door (below)—where unbeknownst to the couple, insectlike creatures are all huddled up in the dark. The trespassing couple awakens the Geonosians—who pursue them into the droid factory. The sleeping quarters were extended with a model (opposite, at ILM; right, final frames).

## Droid Factory

What appears in the movie as a big complicated action scene began as nothing at all. George always finds it amusing to point out that we were able to shoot all the plates we needed for this droid factory scene in three hours. During pickups on Wednesday, March 28, 2001—because one piece of bluescreen conveyor belt was as good as another—we just blew through filming by lunchtime (opposite with stunt Padmé and a bluesuited stuntman; in this shot, Padmé was going to fall off the conveyor belt onto some boxes). We then took the footage of Natalie Portman and Hayden Christensen, and combined it with miniatures and CG elements (right, final frames). The miniature we built to complete the droid factory was one of those *large* miniatures. (Below, scissor lifts were needed to reach some parts of the model; bottom right, Bob Hill shoots the miniature.)

## Execution Cart

After being captured in the droid factory by Jango Fett and sentenced to death by Poggle the Lesser, Anakin and Padmé are sent to an antechamber and put in an execution cart before being towed into the arena to join Obi-Wan (below, final frame). The antechamber set was also used as Dooku's secret hangar. The same archway (opposite, on left of image) that leads out to the arena leads into the secret hangar, redressed of course. We shot the execution cart scene on Thursday, August 3, 2000, and the secret hangar scenes from Monday, August 21 to Wednesday, August 23. Not surprisingly, during the earlier shoot on Stage 2, the art department hadn't finished constructing the solar sailer starship, whose naked hull resides just off camera (opposite, the gray form in the center of the image).

## Execution Arena 1

The execution arena is where the climax of *Attack of the Clones* begins. Creatures arrive to devour the heroes, who are manacled to columns, but their execution turns into a fierce battle between the Republic clone trooper army and the Separatist droid army. In Sydney, Australia, the arena set was sprinkled with packed, crushed stone and adorned with three partially built columns; most of the principal photography of Hayden Christensen, Natalie Portman, and Ewan McGregor was completed on this set (bottom right).

To complement the myriad shots and scenes, we did an extensive shoot with a big $\frac{1}{72}$-scale miniature at ILM. The model was built so that it would come apart in forty-five-degree wedges. We mapped out the direction the camera was looking in each sequence and carefully organized a plan so all shots looking in a similar direction were grouped together. We then formed three different camera crews—because we had about three hundred shots to complete in not much time. We split the model in half (opposite) so two crews could operate simultaneously, with the third crew working at night (top right, final frame with model in background).

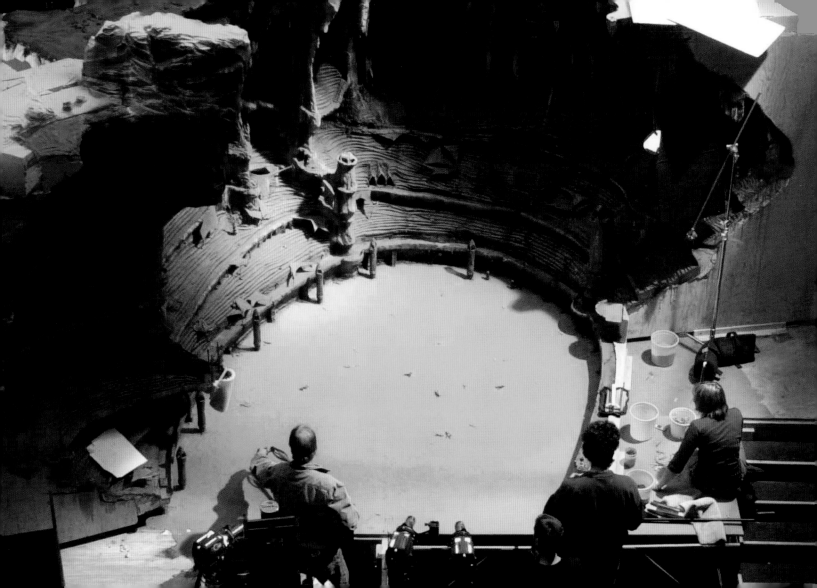

## Execution Arena 2

The first few scenes in the execution arena involve three creatures—the reek, acklay, and nexu—who are out to kill and presumably feast on Anakin, Padmé, and Obi-Wan. Though we had three monsters, we built only one "hero" gate at ILM for their respective entrances—a highly detailed miniature (opposite). In the movie it looks like we have three gates, because we simply dressed the one archway three different ways (top right, final frame). One is in sunlight, another in shadow; the third has torch holders on either side (right, each creature emerging through its redressed gate).

We built some larger-scale sections of the execution arena for crowd shots and for all the close-up shots of the Jedi igniting their lightsabers. The bleacher section and staircase, along with a couple of landings, were redressed to simulate dozens.

The arena crowd is a mixture of CG droids and ILM extras. For the supplementary Jedi, we held tryouts for ILM staff with martial arts experience. They were told to create minute-long routines; only a dozen or so made it into the movie. We shot them from various angles and in different lightings. From their minute-long routines, we took thirty-second chunks; Rob Coleman and his team would then animate to that chunk. Next the droid was rendered out, and a 2D image of both the Jedi and the droid would be put on a card. We'd then use a software tool to help us fill a particular scene. We could program in how many extra cards we'd need; the program would orient the cards toward camera, and choose specific cycles so the action wouldn't be repeated. (Below, final frame of the crowd and Jedi with model.)

## Dooku's Arena Box

First the Sydney, Australia art department built Dooku's full-sized box—from where he watches the arena action—as originally designed (right, on set). In postproduction, however, George decided that it looked too much like Gunga City; it's true that the "melted-candle effect" was pretty characteristic of that Episode I environment. Therefore we junked all the photography of that set, redesigned it, and rebuilt it as a ¼-scale miniature (opposite, with ILM camera assistant Dennis Rogers). ILM crew then had to roto out the occupants of the first full-sized box and composite them into plates of the new miniature box (below, final frame).

## Riding the Reek

A panorama taken in Sydney, Australia shows the actors (Natalie Portman and Hayden Christensen) sitting on a blue hydraulic gimbal (opposite), which was later replaced with a CG reek (below, final frame).

## Arena Battle 1

The dirt floor and partially built columns also formed the set where all of the arena's hand-to-hand combat was shot (opposite, on set): Mace versus Jango, Obi-Wan versus acklay, etcetera. For each shot, the columns were either moved or eliminated as the angle decreed (right, final frames). Posed on the ground, the boxes wrapped in blue fabric (opposite, in background) acted as stand-ins for the floors of the gunships, which arrive like the cavalry, coming to rescue the remaining Jedi. The actors could jump onto these boxes, as if they were jumping into the craft; the actual gunships were added in postproduction (below, final frame).

## Arena Battle 2

Stunt coordinator Nick Gillard rehearses with Hayden Christensen and Ewan McGregor in the arena, which is just a dirt floor surrounded by bluescreen (opposite; below, final frame).

## Deleted Jedi Battle

In the shooting script, while Mace Windu confronts Count Dooku, Ki-Adi-Mundi (Silas Carson) leads an attack on a droid control ship in an attempt to knock out the droid army, as in Episode I. He succeeds in overrunning the ship, but the droids in the arena keep fighting because there's a backup system this time around. George shot the scenes in Sydney, Australia (opposite and below), but they were cut in postproduction.

■ 00114938

## Geonosis War Room

Shot during the second pickups at Ealing Studios in November 2001, the scene in the war room reveals that while the Geonosians may be losing the battle, they have a secret weapon designed to win the war. This sequence was added because George wanted to slip the Death Star into the film, and because he started wondering about what Count Dooku is doing from the moment the clone troopers arrive until his flight to the secret hangar. The war room scene solved both problems: the Geonosians hand over the Death Star plans to Dooku, whose interim whereabouts are now made plain to the audience. Christopher Lee was shot against bluescreen and the environment's reality was filled in with matte paintings and miniatures (below, final frames; opposite, a miniature).

## Attack Gunship 1

On a Fox Studios soundstage, the art department built a partial gunship, with just a hint of the interior and a little bit of exterior outside the door (opposite, on set; right, final frames with extended CG interiors).

Added dialogue—an argument between Anakin and Obi-Wan inside the gunship—forced us to improvise in postproduction, because suddenly there were more than a dozen shots that needed detailed backgrounds. Originally, the CG gunship interior wasn't meant to be full-screen; it was meant to hold up only during brief flybys through the frame. The textures weren't really high enough resolution, and the geometry wasn't detailed enough to withstand the kind of scrutiny a camera inside the ship would demand. We decided to solve our problem by "up-rezzing" the CG interiors. For each shot, we put the camera where it had to be and then the viewpaint and CG departments would work together to upgrade either the geometry or the painted textures, or both. The gunship cockpit was a combination of CG and a miniature (top right, final frame).

## Attack Gunship 2

A panorama taken from the interior of the gunship set at Fox Studios in Sydney, Australia (opposite; below, final frame).

■ 00111524

## Forward Command Center

The forward command center, where Yoda disembarks to monitor the first Clone War, was partially a scale model of the control panels and the interior portion in which the troopers are seated (opposite, with ILM model-makers Dave Fogler and J. P. Kelly; right, final frame). For the wide shot of the forward command center exterior, a CG model was created (below, final frame).

## Geonosis Dunes

Many of the shots in which assorted vessels are cruising over the dunes made use of a miniature consisting of a huge load of dirt spread across an approximately 40 x 40-foot tabletop (opposite—with model-makers Tony Preciado and Nick Bogle—and right, at ILM; below, final frames with the miniature and a matte painting for the sky). Once the shots were finished, model shop supervisor Brian Gernand would have the tons of sand loaded into "dirt boxes"—dumpsters for dirt—and shipped off.

# Dooku's Secret Hangar 1

What was a dirt floor when this set was being used as the execution arena antechamber became a finished piece for the secret hangar floor; the solar sailer was finished, too (below, under construction; opposite, the finished set). Fortunately the majority of this scene was complete in camera, with bluescreen used only outside the doorway and for a couple of the walls (bottom right, a final frame in which a CG ceiling has been added; top right, the walls have been extended with computer graphics). It's here that consecutive duels occur between Count Dooku, Anakin, Obi-Wan, and Yoda.

When George added some new shots in the pickups—close-ups of Count Dooku — we rummaged through stills of the set to find matching backgrounds. Working with George, it's pretty much a guarantee that some sets will need to be reconstructed in postproduction, so I'd taken a number of reference photographs, covering the secret hangar set as well as I could, so we were able to compensate for the pickups without too much extra work.

# 272

## Dooku's Secret Hangar 2

A panorama of the hangar set seen from ground level (opposite; below, final frame).

■ 00106972

## 273

## Dooku's Solar Sailer

The solar sailer, which Count Dooku uses to escape from Geonosis, was a full-sized set in Sydney, Australia (opposite, the ship in upper and lower halves; below, under construction) and a CG model at ILM. George changed the position of the cockpit, however, in postproduction to solve a logistical problem. There was an issue of how Count Dooku was supposed to get into the cockpit. After walking up the ramp into the belly of the sailer, as originally designed, he would have had to crawl through an impossibly tiny part of the ship to remerge inside the cockpit bulb. Therefore toward the very end of production, the cockpit was moved backward until it butted up against the hull and provided a logical entrance, which is what we see in the final movie (bottom right, the chair used for the Ealing Studios pickups; right, final frames).

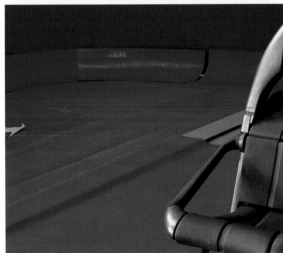

## 274

## Coruscant Secret Hangar

When Count Dooku arrives on Coruscant, he lands his solar sailer in a big, abandoned, run-down industrial hangar. Like the *Millennium Falcon* before it, the solar sailer set-piece was too big to move to another stage, so the day after shooting the Dooku secret hangar scene—Thursday, August 24, 2000—the crew raised a bluescreen to cover its walls and changed the floor tiles to transform Geonosis into Coruscant (opposite, on set). To complete the shot, we then comped the ship and the principals into a miniature of the Coruscant secret hangar built back at ILM (right—note that we replaced all the floor tiles; below, final frame).

## 275

## The Clone Wars

At the end of *Attack of the Clones*, Palpatine and his minions watch as tens of thousands of troopers are loaded into Star Destroyers and take off to fight the Separatists in what is known in *Star Wars* lore as the Clone Wars. Palpatine's perch (opposite, with stand-ins and crew, as we were setting up) was in reality the balcony of Padmé's apartment (below). When they struck the apartment set on July 5, 2000, they tore the whole balcony off—they just ripped it from the set, repainting it and changing the railing a little bit for the shoot on July 13. A matte painting filled in the background (right, final frames).

## 276

## A Secret Wedding

The day before we were to film, we showed up at this site on Lake Como, Italy, in order to scout out the location. All was well—though there were some ominous clouds. The day of the shoot—Thursday, August 31, 2000—it was pouring rain. But George didn't want to stop. He just said, "This must be what it's supposed to look like." In fact, the first scene on the schedule wasn't going to be a problem: in it, Anakin is awakened by a nightmare and goes to a covered balcony to meditate. But we were really worried about that afternoon's two scenes, one of which was to be the secret—and outdoor—marriage of Padmé and Anakin. It wasn't going to be very romantic if they were both soaked. The day of their betrothal, we woke up and it was raining—but, miraculously, just as the crew was setting up, it stopped. About fifteen minutes before the cameras were to roll, the sky actually cleared and we had sunlight. The sun was right on cue, and the clouds had a dramatic, romantic post-storm look (opposite; below, final frame).

STAR WARS: EPISODE III

# BRAVE VIRTUAL
# WORLDS

## Getting Better All the Time

For our second all-digital *Star Wars* film, we were using the same basic technology, but we had the next generation of cameras, which were better in every respect—better signal-to-noise ratio—therefore we got a cleaner picture. We were also able to acquire higher bandwidth recorders, so we had less compression, which meant, again, a better image. We were also recording in 12-bit instead of 8-bit, which made for improved bluescreen extraction. In addition, there used to be a YUV color-space conversion in the camera: the chip was actually recording RGB, but then it was being converted to YUV; when we loaded it online, we had to convert it back to RGB. So there were actually two passes through a color-space conversion—but on Episode III, it was straight RGB to RGB, and that really made the images look better. (Opposite, cast and crew on set at Fox Studios in Sydney, Australia.)

The images from the high definition (HD) cameras were so sharp we had to put Pro-Mist filters on them. In fact, we'd discovered pretty early on while shooting Episode II digitally that things you could get away with on film—like how much makeup actors had on, and the level of finish on sets—would appear fake because the images were too crisp; you could see too much. The HD image didn't jump around—it wasn't as noisy—it was sharper and clearer. Basically, we found that the images we'd recorded on film for Episode I were noticeably softer than what we were getting from the HD cameras on Episodes II and III, so we actually had to "dumb it down" as George says. In some ways the Pro-Mist filters were simply the continuation of a photographic style, since we had previously used them on Episode I; although so much has been printed about digital cameras not having the same resolution as film—it's just not true. The thing is it's not the number of pixels or the theoretical number of resolvable line pairs that optics can give you on film—where under laboratory conditions, one can achieve a certain high level of resolution—what matters is real-world performance. And there the resolution of HD is comparable or better than that of film.

*Revenge of the Sith* (2005)

DIRECTOR: George Lucas
PRODUCER: Rick McCallum
STORY AND SCREENPLAY: George Lucas
PRODUCTION DESIGNER: Gavin Bocquet
DIRECTOR OF PHOTOGRAPHY: David Tattersall
VISUAL EFFECTS SUPERVISORS: John Knoll, Roger Guyett
CONCEPT DESIGN SUPERVISORS: Erik Tiemens, Ryan Church
SUPERVISING ART DIRECTOR: Peter Russell
SET DECORATOR: Richard Roberts
EDITORS: Ben Burtt, Roger Barton
DIGITAL MATTE SEQUENCE SUPERVISOR: Jonathan Harb
MODEL-SHOP SUPERVISOR: Brian Gernand

## 279

## Opening Space Battle

While the differences between principal photography on Episode I and II were so great—new country (Australia), new crew (though the department heads were mostly the same), new recording format (digital)—Episode III had a familiar feel, because nearly nothing dramatic had changed since Episode II. We had pretty much the same crew, and we were in the same country shooting at the same studio on the same stages. As of our first day of shooting, on Monday, June 30, 2003, it had been three years since *Attack of the Clones*, but it felt like we'd just taken a short break. We all fell right back into the same pattern after a week or two of filming.

For the opening space battle, in which Anakin Skywalker (Hayden Christensen) and Obi-Wan Kenobi (Ewan McGregor) have an aerial dogfight with droid ships while trying to rescue the Chancellor (top right, final frame), we had two versions of the Jedi starfighter: a cockpit and a fully dressed ship (middle right: we repainted the exterior for pickups of their respective starfighters; all the point of view shots were done at ILM, as well as the moment when, bottom right, Obi-Wan's astromech is beheaded). In Sydney, Australia, the cockpit portion of the space battle was shot in one day. Hayden and Ewan were recorded in the same set, because there was only one. It was redressed a little bit for its respective occupants: After Ewan had completed his lines, the art department drilled some holes and put some LED lights on the bulkhead behind Hayden's head. One problem with the cockpit, which I found funny, was that when we were doing close-ups of Hayden, we discovered that if he put his feet on the actual pedals in the cockpit, his knees raised up into the shot—and we didn't want his knees to block the frame. Consequently, he had to stick his feet out the end of the cockpit (opposite page).

## 280

## Trade Federation Ship Hangar

The Jedi manage to land in the hangar of the Trade Federation Cruiser, where the Chancellor is being held hostage by Count Dooku and General Grievous. There were three versions of this hangar: just a floor that we shot in Sydney, Australia (top right, with stunt coordinator Nick Gillard); a miniature that represented ¼ of the hangar (opposite page, with ILM model maker Grant Imahara); and what we called, "the Artoo wall," a much larger ¼-scale "hero" portion of the set for close-ups (below right, a final frame combines all three versions: on the right is the Artoo wall; the principals are on the Sydney set; and the background makes use of the ¼-scale miniature).

Putting all these versions together, we decided to conduct a bit of an experiment. One of the big concerns is always cost, and we calculated that it would be about the same to build the miniature or the CG model. The major difference was going to be the shooting time: Each day you're out on the stage shooting a miniature is expensive. So we asked ourselves a question: What happens if we build the set as a fairly detailed miniature, take some really nice high-resolution stills of it, build a CG model of the basic shapes with simplified geometry—and then stick the photographs of the miniature onto that geometry, creating a "Zenviro" version of it? We tried this and it worked out well (below left, the Zenviro version). With this we eliminated much of the costs, and enabled simultaneous shooting.

## Trade Federation Ship Hallway

We built a set piece in Sydney, Australia, of the ship hallway—but of course the hallway was meant to look much longer. This was another good case for Zenviro, so I thoroughly photographed the set; later we built relatively simple geometry that those pictures could be stuck onto. In this case we Zenviroed a full-sized set (below, a final frame of the destroyer droids rolling up the Zenviro hallway to blast away at the Jedi as they enter the elevator). In the shooting script, R2-D2 tries to follow the Jedi down the same corridor on their mission, but Anakin tells the droid to stay behind; later this was changed so that R2-D2 never leaves the hangar (opposite, the droid "waiting" for his ultimately deleted hallway scene to be shot).

## The Deaths of Shaak Ti

The shooting script had the Jedi following Shaak Ti's (Orli Shoshan) homing beacon. They find her a prisoner of General Grievous, who kills the Jedi Master aboard the Trade Federation cruiser. The scene was shot with Duncan Young as a stand-in for the general (opposite). Her death was moved to the Jedi Temple (below), during pickups at Shepperton Studios, where Anakin was to stab her, but this scene was eventually cut from the final movie.

## Fuel Tank

Once again the shooting script had scenes that didn't make it to the final movie (though some of them found their way to the DVD). Obi-Wan and Anakin are surrounded by dozens of droids while Shaak Ti is killed, so they cut a hole through the floor with their lightsabers and fall into the ship's fuel tank. Some of the shots here had the actors knee deep in the tank against bluescreen, sloshing around in the water (opposite). The Jedi continue their escape, but the fuel tank level rises as the droids pursue them. The art department therefore built a set piece of a wall, which was on a winch, so the whole thing could be raised and lowered (top right) to simulate whatever height the water level needed to be (bottom right, a rehearsal for the moment in which the water is up to their necks).

    To complete the sequence, I had planned on doing a miniature, but it would have been extremely tricky and big and expensive—so we weren't very disappointed when this part of the movie was cut in postproduction. George deleted it because the whole beginning—the opening space battle, the cliffhangers—was intended to be a half hour long, but was turning into the first hour and twenty minutes of the movie.

## Vent Shaft

Finally Obi-Wan and Anakin arrive at the end of the fuel tank. Anakin senses an opening behind a grating in the ceiling, pulls it off, and they climb up into a vent shaft.

The vent shaft was built as two sets in Sydney, Australia: the big area (right) and the narrow area (opposite). There was actually a pretty cool rig for the latter set. The small shaft was only two feet in diameter, so a person could barely get through. Therefore, they set up a camera on a "Rhone-truss" (a reenforced sixteen- to twenty-foot-long lighting tube) and stuck the camera on the far end of it; the other end was attached to the dolly with a lot of counterweight on it. This way, as someone crawled forward, the crew could dolly back with the camera well within the confines of the narrow shaft.

## Elevator Lobby

Here again is a set that had multiple uses: The lobby where the Jedi get an elevator up to the general's quarters (opposite, on set) is the same lobby they exit into once they've arrived (below, final frame; note that we replaced one wall with a window, through which one can see part of the space battle). Later, during their escape with the Chancellor, they swing out of the elevator shaft into the same set (bottom right, final frame). Plus, for easier access and lighting, one elevator interior was detachable, so we could do scenes like the one in which Anakin and Obi-Wan are surrounded by battle droids (top right, final frame).

## General's Quarters 1

The Jedi find and liberate the Supreme Chancellor, then do battle with Count Dooku in the general's quarters. It took a lot of planning and several weeks to build this great big set that filled up half of Stage 7, which is reminiscent of the Emperor's throne room in *Return of the Jedi* (below).

Nevertheless it lacked a ceiling and a few walls. To compensate, we built a CG model (top right, plain CG model) and rendered it: adding texture, color, and lighting (bottom right, completed CG render). Almost every final frame of this environment uses some part of the render, because everywhere the camera looked, the set was missing elements. The spaces between the stair steps were a particular challenge (opposite).

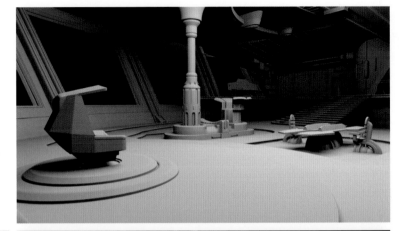

## 287

## General's Quarters 2

The other big issue for these scenes in the general's quarters (opposite) was actor Christopher Lee's schedule. He plays Count Dooku, who nearly does away with Obi-Wan but is beheaded by Anakin. In the summer of 2003, Lee had flown to New Zealand to do some pickups for *The Lord of the Rings: The Return of the King*. Flying all the way around the world from his home in England is a lengthy and stressful thing to do, and no one wanted to ask him to make a return trip to Australia. Therefore an effort was made to schedule a day with us for his *Revenge of the Sith* role right after his New Zealand trip. The problem was that the set where Count Dooku has his scenes—the general's quarters—wasn't even scheduled to start construction until after Lee's return to England. Therefore we had to shoot all of his elements bluescreen—which meant that director of photography David Tattersall had to commit to a lighting scheme before he'd actually lit the real set. We came up with an idea of the grids in the floor and ceiling and, looking at a maquette of the future set, George decided where Dooku would say each line. As a result, anytime you see Christopher Lee, it's a CG set behind him (below, final frames).

## 288

## Elevator Shaft

We built a little section of the elevator shaft in Sydney, Australia, for when Anakin is left hanging on a ledge, after being stranded by a wayward elevator (opposite, preparing for a shot of Anakin on top of the elevator roof, before it descends without him). We also used this portion of set for when he's hanging there with Obi-Wan on his back and the Chancellor clinging to his foot. To extend the set, we built a CG model of the whole shaft (top right, untextured geometry; bottom right, completed CG render; below, final frame).

## Elevator Revolve

Located on Stage 1, the revolving rig was used for many partial sets, including the Trade Federation bridge, elevator, and elevator shaft. For the shots of when the ship is falling out of orbit, and the elevator lobby tilts as the trio are waiting for the elevator, the art department created a two-sided set within the revolving rig: lobby on one side, and shaft on the other (opposite and below left; top right, final frame). We also used the revolving set for when Anakin is hanging in the elevator shaft with Obi-Wan and the Chancellor (bottom right). Because the shot was fairly wide, we extended it with a CG model (middle right, final frame).

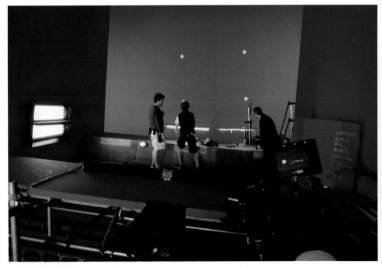

## Editorial

Next to our makeshift ILM office on the second floor of Stage 2, editor Roger Barton works in his glamorous editorial suite. Thanks to the immediacy of the digital medium, Roger was able to create a rough assembly of the movie as we shot each scene.

## Ray-Shield Corridor

The art department built a corridor set for the scene in which Obi-Wan, Anakin, and the Chancellor are trapped in a ray-shield prison. Mostly complete in camera, this set was also used with a different lighting scheme for when Shaak Ti is killed (opposite, lighting for Shaak Ti's death; top right, final frame of the trio caught in the ray shield; note the different lighting and color schemes).

To save time and money, we didn't shoot plates for any of the reverses; we recorded only the principals. So when destroyer droids roll in, or when R2-D2 arrives, you're looking at a Zenviro version of the set, which was digitally reconstructed from about 200 photos (bottom right, plain CG set; below, textured CG set). As written, the ray-shield scene was longer than what ended up in the movie, and I was sorry to learn that one bit of humor was cut. Originally, the trio discuss how they might escape, and Palpatine has a suggestion: surrender and work out a negotiation. He gives his whole pitch to the Jedi and, once he's finished his speech, Anakin and Obi-Wan turn and look at each other as if nothing had been said.

## Trade Federation Bridge

This bridge was a big elaborate set, which was still missing a ceiling, the end of the hallway, and other elements (opposite and bottom right; below, underneath the set is a forest of pipes and lights with green gels). The trio is taken here to be questioned by General Grievous.

George likes to shoot really fast—the number of setups we did per day was really high—so when the time we were going to be on each set was first scheduled, there was a lot of optimism concerning how quickly we were going to blow through this stuff. Usually the schedule was on target, but for the bridge scenes, we didn't allow enough time. I'd asked beforehand if we could shoot the General Grievous side of the set, so we'd have all the reverse shots of him and his CG bodyguards; I'd wanted to shoot a stuntperson in a blue bodysuit and empty plates for reference. But when we started running behind schedule, George jettisoned all of that and said, "You guys are going to have to come up with all of those backgrounds because I don't have time to shoot them." While he completed the scenes with the principals, I went to extra lengths to cover the bridge, taking still photographs of absolutely everything. Later at ILM, we built a CG model—so anything on the bridge with a bodyguard or General Grievous is a virtual set (top right, final frame).

## Rotating Bridge

The Trade Federation bridge scenes originally consisted of a few shots in which the Jedi fought with droids while the ship revolved. The droids magnetize while the Jedi are tossed around, still cutting down their adversaries. For these sequences, the art department first created a revolve with a complete ceiling set (bottom right)—but that was later changed to a ceiling with blue shapes, because it was felt that would be safer for the actors. We did shoot the rig turning with the principals hanging and fighting battle droids (opposite and below), but all of the bridge revolve was cut from the final movie (top right, final frame: only General Grievous and the droids are on the bridge when it rotates in the final cut).

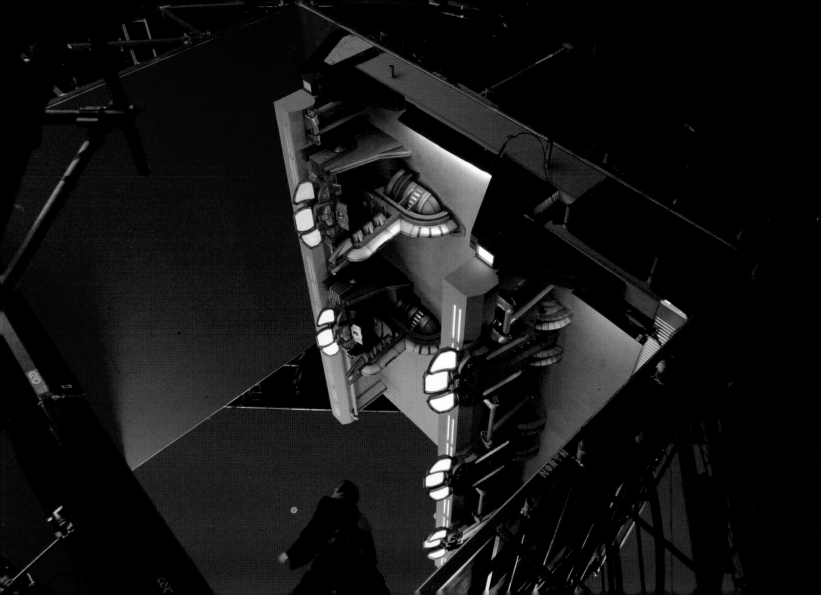

## 294

## Escape-Pod Bay

The escape-pod bay wasn't built as a set on a soundstage, because the only character in it was computer generated: General Grievous. The escape-pod bay was the very first CG environment we started working on for *Revenge of the Sith*, so this was a bit of a test case for us. Not only had George requested that we do more CG environments than on Episode II, but I'd also made the argument that if we were going to create more of these environments, we needed to find a nice way to get radiosity into our "pipeline" (ILM's internal production structure)—because CG interior environments are so dependent upon diffuse inter-reflectivity, or "radiosity." Real-life interior environments often have very little direct illumination: instead light is bouncing off things, and much of the illumination might be second or third or fourth bounce, not from a direct light source. Interior environments are completely dominated by this effect, so if you want to do a CG room realistically you have to have lots of diffuse inter-reflectivity.

It's only in the last few years that computers have become fast enough to be able to calculate radiosity. The idea started ten years ago, but it would take a week to render a simple low-resolution interior back then. At ILM, we like to keep our renders down below four hours a frame. For the escape-pod bay we were able to set up the lighting once, take the hit for the long calculation, and then "bake" that into the surfaces (opposite, the gray, untextured version of the CG model). Normally most of the lighting in computer graphics is calculated on a per-frame basis; because the objects move around, you can't do it all in one go. In the case of environments that aren't moving around, a lot of the expensive math can be done just once and then "baked" into the environment for subsequent frames. So we built this whole interior set and placed the lights on the ceilings; after that, we could render as many frames as we wanted, with some nice diffuse bounce that worked out pretty well (right, final frames).

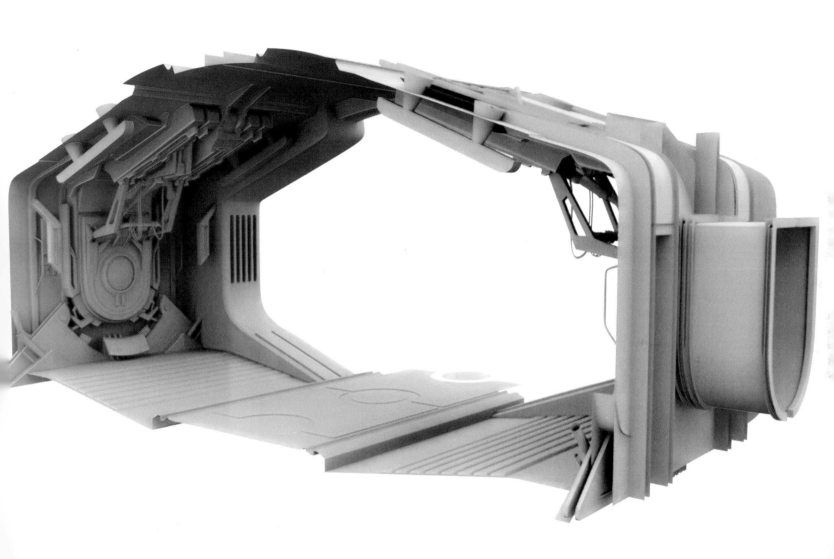

## 295

## Senate Panorama

A global illumination render of the geometry created by digi-matte painter Yusei Uesugi in preparation for his digi-matte painting of the Senate and its ancillary office buildings (opposite; below, final frame).

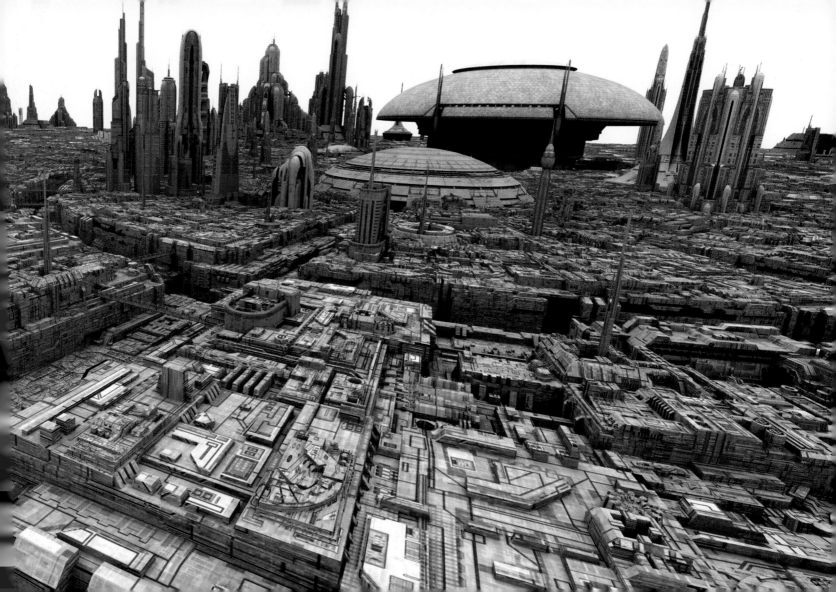

# Senate Landing Platform

Shortly after Anakin has managed to crash-land the Trade Federation cruiser, the trio is greeted by Mace Windu (Samuel L. Jackson) and assorted dignitaries at the Senate. In Sydney, Australia, we had a partial interior of the shuttle, a section of the floor, and a couple of partially built columns. On this set, as on many of our sets, DP David Tattersall would lay in a whole grid of soft lights, each inside a cylindrical translucent housing, like a Chinese lantern. Six feet below that grid, they'd extend a big silk sheet that would further diffuse the light, spreading it around to create a soft effect (opposite)—which is a really good way to simulate a daytime exterior environment. Then he'd put in a key light, which is the hard light, to represent the sun.

Because Obi-Wan is standing in the doorway in this scene and he was going to be seen in close-up, I wanted to have an interior of the ship. Production designer Gavin Bocquet obliged, which meant I had a background for all those shots. But George decided in postproduction that he wanted to take the ship and flip it around; he felt that what we had modeled as the back of the ship should be the front, which meant that I couldn't use the interior anymore. All of that ship set was therefore scrapped, and we ended up rotoing around Obi-Wan—cutting him out—and placing him next to a CG ship (below left, final frame; below right, the wide shot was completed with a digi-matte painting).

## Senate Hallway

As the politicians walk toward the Senate interior, Anakin leaves the group to find his wife, Padmé (Natalie Portman; below, final frames). They embrace in the hallway, whose multiple columns originally were neither so high nor so numerous (opposite, on set at Fox Studios).

## Padmé's Veranda

In Episode II, we shot a scene with a walk-in closet in Padmé's apartment (see spread 195)—but we only saw part of it. When we returned to this set in Episode III, George had us go into that undefined portion (this is something he does a lot, which I think is kind of amusing). Therefore the walk-in closet became part of a corridor with a few steps that lead to Padmé's veranda (top right).

This was a huge, sometimes complete-in-camera set, but as the veranda was often seen in the context of Coruscant, several digital matte paintings were created for different times of day (middle and bottom right, final frames with a digi-matte painting by Giles Hancock). The veranda was a huge job for the Sydney art department, and one question we had before starting was how much of the set to build. I always ask George and Rick McCallum for more set (ILM always has plenty to do, so it reduces our workload). I also always push for having a floor, because someone standing on a digital surface is still hard to create; it's difficult to get the reflections and shadows just right. Ultimately, the veranda set featured a terrific floor section built by the art department, and many of the shots were facilitated because of it (opposite).

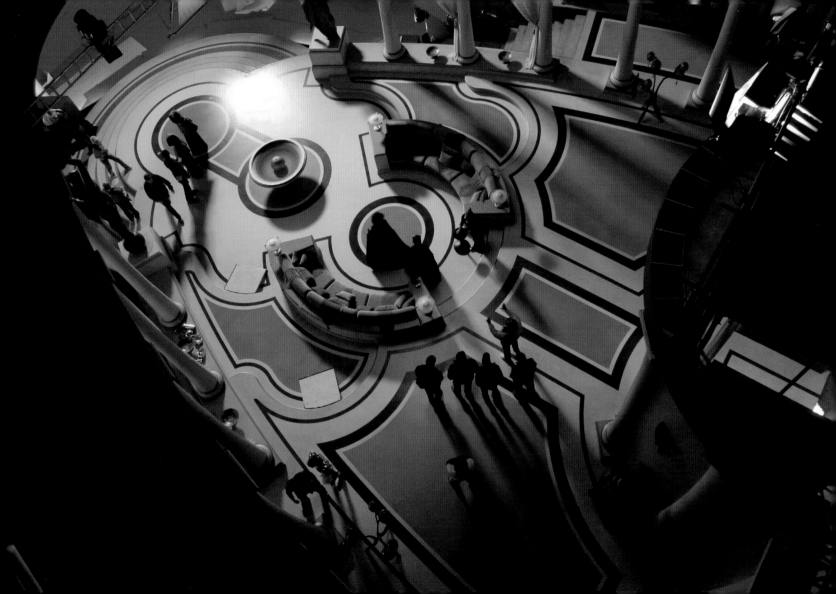

## Apartment Building Model

A panorama of ILM crew working on a shot of the Padmé apartment building miniature (opposite; below, final frame).

## Yoda's Meditation Chamber

One of the two scenes in Yoda's meditation chamber—another minimalist set—was cut out of the movie. In it, Mace, Obi-Wan, and Yoda (voiced by Frank Oz) discussed the bad situation of the Republic, and the dark shadow of the Chancellor (opposite, with stand-ins; top right, deleted final frame). The other scene in this locale was shot during the pickups at Shepperton Studios in England. In it, Anakin tells Yoda about his nightmares, and the Jedi Master explains the concept of greed (below, final frames).

## Palpatine's Office

Because of the many scenes that take place here—meetings, revelations, seductions, and deaths—Palpatine's office complex became increasingly roomy in Episode III (a scene in the reception room outside the office, where Mace Windu and his fellow Jedi make short work of some guards and Dur Wac, was cut). George decided that the 90-degree arc of wall we didn't build on Episode II for camera access (opposite in center of image; bottom right, on right) was in fact another doorway, with a corridor to a private office (also called the antechamber). Once again, George said, "If nothing has been established, I can put something new there."

The office complex was one whole set, but the new part is much grayer than the brilliantly red office (top right, final frame). George mentioned to me that the color red has historically been used as a show of wealth and power. It used to be that only royalty could afford bright red clothing because the dyes were so expensive, hard to obtain, and labor intensive. It's also the color of primal rage. So you have a man with this internal rage—Chancellor Palpatine, aka Sith Lord Darth Sidious—but as his power grows the red fades out, and his world becomes gray. He's becoming completely cold and mechanical, much like the Death Star. The passion goes away, and he achieves a sort of Zen-like evil.

## Palpatine's Private Office

Starting with Episode III, for each set we shot on, George would do a "control take"—that is, the two digital cameras would record the angle at which we were going to shoot the actors, only without the actors. The control takes gave us blank plates and great reference—and were the result of a learning process that began on Episode I.

For *The Phantom Menace*, I shot a lot of general reference with two film cameras. We ended up using them for visual effects plates and all kinds of things—but I wanted to make sure that on Episode II we were getting better coverage. I became more systematic and tried to anticipate what George might be adding. Nevertheless, a whole new scene in Palpatine's office with the Chancellor and Anakin was added late in postproduction, and was shot against bluescreen—so we had to come up with all the red-room plates. We ended up taking a different scene that we'd shot in the office that was filled with people—Jedi, Senators, et al. We cleaned them out, and made floor and wall patches; in some shots of the new scene, we were able to use reference photography, but camera moves needed running footage. The whole process of re-creating the set became expensive and painful.

I think that scene in Episode II was what motivated George to start doing the control takes on *Revenge of the Sith*—and it worked out really well. When he wanted to add a shot of Palpatine in his private office (opposite), we shot it bluescreen the first day of pickups at Shepperton Studios. We then took that plate of Ian McDiarmid and added it to the control take recorded at Fox Studios in Sydney, Australia—and it worked out perfectly. You just can't tell that Ian wasn't on the real set (below, final frame).

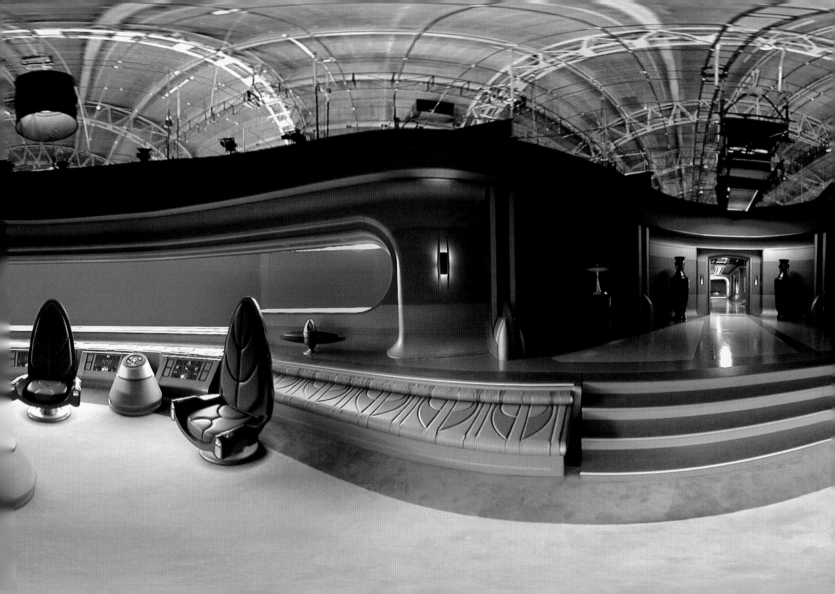

## Jedi Briefing Room

The shots were designed to be complete in the Jedi briefing room (bottom right, a maquette shows how much of the room was actually built to accommodate the desired angles), where Anakin and Obi-Wan meet and talk (opposite, George discusses the scene with Hayden Christensen and Ewan McGregor). For a wide shot we had to build the other half of the room and add map graphics (below, final frame); for a shot of the Jedi exiting, we had to build a CG hallway outside the briefing room.

# Jedi Temple Hallway

After a Jedi Council meeting, Anakin and Obi-Wan walk down a flight of steps into a windowed hallway. The set was of course partial (opposite), while in the movie, outside the vertical openings, is a big 3D digital matte painting by Jeremy Cook. In fact, on Episode III, the line between digi-matte painting and computer graphics started to blur. Matte paintings were being done with camera moves, built with 3D geometry, and lit. Nevertheless, they were still being done by matte painters with a matte painter's sensibilities. They would render an image (bottom right, Cook's global illumination model), paint on it, and then project it back onto the geometry so it would move in 3D (top and middle right, final frames). Ultimately, digital matte painting is becoming a complex mixture of 2D and 3D.

## Opera Hallway

Since Episode I, I've been trying to persuade George to make a cameo appearance. A couple of times there were opportunities: He could have been one of the dignitaries at the Naboo peace celebration, but he thought it was too cheesy. I tried once more on Episode II, and failed again. So I gave up. It took his daughter Katie to convince him to join the opera crowd in *Revenge of the Sith*. Fairly late in postproduction, concept artist Sang Jun Lee, with George's feedback, created a costume, and costume designer Trisha Biggar made it real. We shot the hallway, extras, and Hayden Christensen against bluescreen during the pickups at Shepperton Studios (below). George (Baron Papanoida) and Katie (Senator Chi Eekway) were recorded later, and composited into the hallway (opposite, final frame—George is on the left of the image; his other daughter, Amanda, is in the middle).

## Palpatine's Private Box

This sequence—in which Palpatine further corrupts Anakin by telling him the story of Darth Plagueis—was originally set in the Chancellor's office, but George was feeling that we'd seen that big red room too often. A lot of scenes had been filmed on that set for Episodes II and III—enough so that George felt he didn't want another long dialogue scene there. So he came up with the idea that Palpatine is out watching a ballet, or something similar; it had to be a form of entertainment that would be relatively quiet so the actors could speak the hushed dialogue of this particular scene.

George changed things around only about two days before we were scheduled to shoot, so there wasn't time to build a set. Instead, Gavin Bocquet's department took four chairs from Palpatine's office and altered them—repainted them—and the rest was bluescreen (bottom right). At ILM, to fill out the locale, we built a miniature, recorded opera extras, and created the Mon Calamari dancers "swimming" in giant liquid orbs (top right, the miniature and model-maker Jon Foreman; opposite, Jason Snell, Carl Miller, and myself looking at a reference print of one of the plates and discussing the spacing between the chairs; below, the opera boxes extended with a digi-matte painting, including the extras shot at ILM).

 307

## Bail Organa's Office

It's really too bad the scene in Bail's office was cut out of the film. It was a really cool set with giant windows (opposite and below); it was also an interesting scene, with the conspirators plotting to form what will become the Rebel Alliance—but that whole plot line was jettisoned because George felt that in order for the story to work, it had to concentrate fully on Anakin.

Organa's office was meant to be symmetrical, but only one wall was built. To complement that one, for the reverse angles, I asked DP David Tattersall to change the lighting in order to simulate the correct sunlight—therefore he reversed the key—and I carefully photographed the existing wall. Later ILM built a Zenviro version of the second wall, which we'd completed before the scene was cut (but which will probably appear on the DVD).

## Jedi Council Chamber

*Revenge of the Sith* necessitated the third and final build of the Jedi Council chamber set (opposite, looking over George's shoulder as they slate the shot), though we'd saved the chairs from the previous film. This time around, the set was used for only one locale (bottom right, production art department maquette of the chamber; below, final frame).

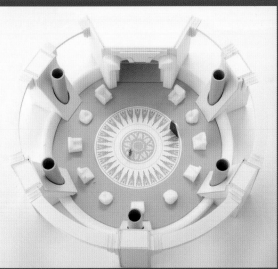

# 309

## Jedi Council Chamber Skyline

There were a few scenes in the Jedi Council chamber, so we needed several really nice digi-matte paintings. Jonathan Harb supervised Yanick Dusseault's cyclorama. The skies were rendered with global illumination techniques and indirect bounce, creating a nice realism in the shadows. George was very specific about what he wanted in the skies throughout the movie, and if you look closely you'll detect very deliberate weather patterns from scene to scene and planet to planet.

# Attack Gunship Interior

Our whole set took up not more than the edge of a converted landing platform; all we did was add a few crates to create the gunship interior (opposite, with stand-in Yoda puppet). We shot the principals against greenscreen, and in postproduction added the necessary CG interior (right, final frames). For the shot where Yoda exits the gunship after it lands, a mobile cutout Yoda stand-in was used for the actors' eyelines (below, Yoda on skateboard).

## Jedi Embarkation Platform

For the scene in which Obi-Wan goes off to assassinate General Grievous, we had pretty much no set at all: just a bit of floor and a railing (opposite, on set; below, final frame). In addition to the massive amounts of animation that Rob Coleman and his team did, the digi-matte department created a painting, and the model shop built a miniature of the embarkation center to complete the scene (bottom right, with ILM effects DP Pat Sweeney, who shot multiple copies of the model and mirrored the move, as the final frame, top right, reveals).

## Jedi Cruiser Hangar

After taking off in the cruiser, but before leaving in his starfighter, Obi-Wan briefs the clone commander on the Utapau mission. In Sydney, Australia, we used the same set floor that we'd used for the Trade Federation cruiser hangar (opposite and bottom right), though later we had to replace it, as the brown tone was too recognizable. Ultimately the whole environment, except for the full-sized starfighter, came from the digi-matte painting department (below, final frame).

Because Obi-Wan's first starfighter was destroyed in the opening space-battle sequence, he's actually using what is supposed to be a new starfighter with a new R2 astromech unit. The G9's copper trimmings (opposite, center of image) were chosen by fans from a choice of four that were posted on starwars.com.

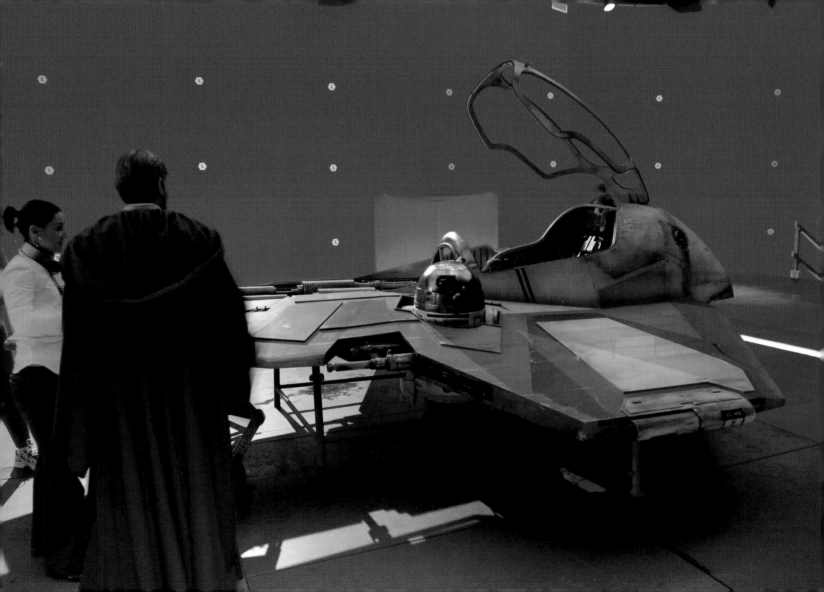

## Palpatine's Private Office Hallway

In this hallway two important things happen: Palpatine reveals to Anakin that he's a Sith Lord, and Mace Windu and Darth Sidious fight.

Originally, instead of bluescreen in the hallway area, the Sydney art department had put in decorative details. But the day before the shoot, George changed his mind and had them pull out those pieces and put in bluescreen, because he decided that he'd rather have an elaborate sculpture in the hallway (opposite). The artwork—which concept design supervisor Erik Tiemens drew and Richard Miller sculpted in postproduction—was about four feet across and tells the story of some ancient historical or legendary battle between aliens and humans (below, the sculpture, in which some of the personages resemble "grays"—the aliens purporting to visit our planet; right, final frames).

## Jedi Temple Hangar

After Palpatine tells Anakin that he's a Sith, Skywalker reports his findings to Mace Windu in the Jedi hangar—a scene that George wrote in postproduction and shot during pickups against bluescreen without any set at all. We recorded it in the sum total of a few minutes, at the very end of the day on Monday, August 30, 2004. George said, "We can't fall behind schedule," so he sneaked it in, and we just powered through it at Shepperton Studios. Besides the principals, the whole set is CG, and it turned out pretty well (opposite, a global illumination render reveals many of the hangar's details that can't be seen in the final frames, below).

# Kashyyyk

The Wookiee homeworld of Kashyyyk (middle right, final frame)—where Yoda travels to prevent a Separatist takeover—was a complex combination of an ILM miniature, plates shot in China and Thailand, digi-matte paintings (below), and Wookiees shot against bluescreen during two days of pickups in Sydney, Australia. The Kashyyyk tree "hero" model was supervised by Brian Gernand and Don Bies (opposite, based on a sculpt by Danny Wagner, which was based on a painting by concept design supervisor Ryan Church). The location plates were shot by Ron Fricke and Carl Miller, and the Wookiees were directed by Rob Coleman on May 17 and 18, 2004, at Fox Studios (top right).

## Kashyyyk Hologram Room

As the plot thickens in *Revenge of the Sith*, the Clone Wars intensify, and the Jedi have a meeting—with some Knights in the Council Chamber, and others on far-flung planets (Silas Carson as Ki-Adi-Mundi and a few other Jedi were recorded at Shepperton Studios in London), while Yoda communicates from the hologram room on Kashyyyk (below, final frame). With him are Chewbacca (Peter Mayhew, reprising his famous Wookiee role) and Tarfful (Michael Kingma), who were shot against bluescreen at yet another location.

The room with the hologram table was built as a ¼-scale model (opposite, with model-maker John Duncan and ILM photographer Marty Rosenberg). As the shot neared completion, George felt the table looked too new and asked that it be dirtied up, so we painted on one frame and projected the paint onto a flat surface that was rendered with the camera move (bottom right, final frame).

## Utapau Sinkhole

We built two sinkhole models, both supervised by Brian Gernand and Mike Lynch: one ¹/₉₀-scale (opposite, for close-ups) and the other ¹/₂₀₀₀-scale (right, for wide shots; below left, final frame). We'd decided to build miniatures because of the mountainous walls, which are hard to do really well in CG, whereas the ILM model crews have loads of experience sculpting foam into good-looking rocks. Nevertheless, it was an expensive build, with all the architecture in there, so we designed the models for maximum "redressability." The same foam block and miniature buildings of the ¹/₉₀-scale sculpt came to represent a lot of different areas: where Obi-Wan first lands; all the backgrounds of the Utapau battle scenes; behind Commander Cody when he hands over the lightsaber; the backgrounds during the scooter/lizard chase, etcetera. There were four major blocks, each with an arched ceiling piece. Plus, the buildings were all detachable, so we shuffled those around; we'd move a building from the background to the foreground, turn it ninety degrees, and light it a different color.

Because there were so many shots, we used the ¹/₂₀₀₀-scale model in a handful of hero shots. Everything else was from a Zenviro version (hundreds of stills were also used). A digi-matte painting by Brett Northcutt was incorporated to create the Utapau establishing shots (below right, final frame).

## Utapau Landing Platform

When Obi-Wan lands on Utapau in search of General Grievous, his landing platform was at first a little section of dressed floor (opposite, at Fox Studios; top right, final frame completed with a CG model and buildings from the ¹/₉₀-scale miniature). When you're looking into the sinkhole crack past Tion Medon (Bruce Spence), the platform and canopy are CG, but what's inside the crack is also all those models (middle right, final frame). The background in the reverse of Obi-Wan is hero photography of the ¹/₂₀₀₀-scale model (bottom right, final frame).

## General Grievous's Chamber

When General Grievous arrives on Utapau in his shuttle, he takes an elevator and repairs to his private chamber to talk with Darth Sidious via a hologram (right, final frames). As with the escape-pod bay, because Grievous was CG, we built an all-CG environment of his secluded abode. Keeping with the idea of using global illumination in all of our environments, his chamber was rendered with diffuse bounce passes. It was a really heavy model (opposite)—it used lots of polygons, geometry, and sophisticated shading models. The per-frame render times for his chamber and other Level 10 interiors were super high—days a frame—so for the vast majority of shots (about a hundred), we were doing one high-resolution frame and then keeping the camera in there, panning and tilting—but we tried to keep big camera moves to a minimum.

## Utapau Level 10

In a conceptualized sequence that was never animated, a Neimoidian core ship detaches from the outer ring, descends to Utapau, embeds itself in a sinkhole wall, and opens up all its doors and extends its antennas. In the movie, we see only the embedded core ship—which is the site of a big battle and where Obi-Wan and General Grievous begin their duel.

We decided to create Level 10 (the core ship interiors) with computer graphics, because the design included so many repeated pieces: Consoles, railings, and so on. When these sequences were first turned over to us, we did what we usually do: We carefully mapped out where each character was, and where each action was taking place, and built the environment to accommodate all of that—a jump here, the scooter parking place here. It was a two-month CG set build, and a heavy model (opposite, the gray, untextured version; middle right, a final frame). But as George was editing, he shortened the sequences.

It was too late to change the CG model, so I used a cheat: When Obi-Wan drops the block onto the magnaguards, he walks by the block to face General Grievous (below left); they begin their fight, and move backward (below middle)—but too fast, because three shots later they're about a hundred feet behind the block. In theory, it should take them longer to get there. Rather than point out what the problem was to George, I just got rid of the block, so the lingering landmark disappears as fast as possible (top and bottom right).

## Utapau Main Plaza

After Obi-Wan makes short work of General Grievous's lightsaber arms, the General flees
on his scooter bike, with the Jedi pursuing on his lizard-mount. A breakneck chase ensues
through the sinkhole interior—for which we used the same $\frac{1}{90}$-scale rock blocks rearranged,
with a chase track for planning (opposite, with camera assistant Dennis Rogers and DP Carl
Miller—the track is on the left of the image). For all the shots in which we're hurtling along,
we did a corresponding sphere pass; that is, we hung the chrome sphere on a post attached to
the camera and shot the sphere for the sequence. We used those shots directly for the render,
because General Grievous's scooter bike is metallic and it needed to reflect the environment; a
computer tool extracted the reflection environments from the sphere, so the running footage
really helped in completing the shots (below and right, final frames).

## 322

## Secret Landing Platform

General Grievous is trying to escape and leads Obi-Wan to his secret landing platform, which is in an adjoining, smaller sinkhole, only about 300 feet in diameter. We shot Ewan McGregor at Fox Studios using the same dressed floor as the other Utapau landing platform, with a partial set piece of Grievous's fighter brought in (opposite). For their fight a stuntman in a blue bodysuit stood in for the CG Grievous.

We didn't create an additional model for the smaller sinkhole. Instead, we took a high-resolution 3D scan of the ¹/₂₀₀₀-scale model and bent it around to make it into the smaller sinkhole—which worked for the wide shots, as the ship leaves (right, final frames). For the rest of the shots during their fight to the death we used stills of the ¹/₉₀-scale model for the background (below, final frames).

## 323

## Utapau Sinkhole Grotto

The sinkhole grotto, into which Obi-Wan and his lizard fall after traitorous clone troopers blast them (right, final frame with a digi-matte painting by Eric Ippen), was originally part of an extended sequence. Obi-Wan was going to emerge from the sinkhole lake and climb up into a cave, where he'd hide while seeker droids sought him out—but these were going to be eaten by a Nos monster, allowing the Jedi to escape.

In Sydney, Australia, we shot Ewan McGregor on a little chunk of dressed rock we submerged into the tank (opposite). We also built a miniature and had a digi-matte painting of the rock for a shot when the camera pans over to the seeker droids (below, final frames stuck together show the pan). We did an additional close-up of Ewan in Shepperton Studios, because we needed a little action of him putting away his breather. The edited version of the film, however, contains only the rock climbing section; the Nos monster didn't make the final cut.

## 324

## Grievous's Starfighter

Based on concept art by Ryan Church, the art department in Sydney, Australia built the cockpit
of General Grievous's starfighter (opposite, with first assistant director Colin Fletcher in
background), which Obi-Wan uses to escape Utapau and rejoin Yoda and Senator Bail Organa
(Jimmy Smits). The cockpit controls (below), though intended for a droid, apparently pose no
problems for the Jedi.

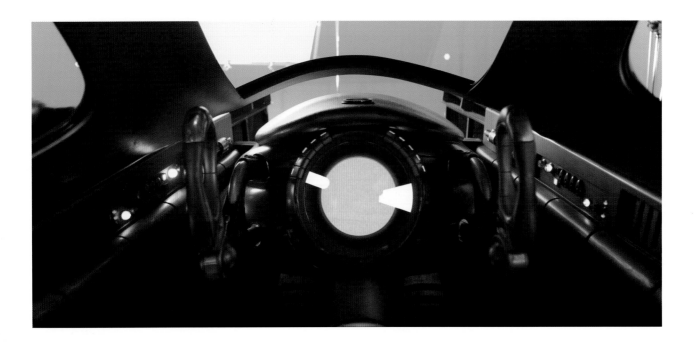

## Felucia

The planet Felucia is featured in a short sequence where Jedi Master Aayla Secura leads a band of clones who turn on her (top right, final frame—Amy Allen reprises her Episode II role). This was a challenge because Felucia was probably the most alien environment in the *Star Wars* saga: really weird plants with a back-lit, transparent look. Allen was shot against bluescreen in postproduction and, while the flora were going to be CG, I asked for a miniature to be built for reference. Once again, I felt like things could go badly if we weren't grounded in some kind of reality—Felucia's odd look could have easily become nonphotographic. We built foot-high maquettes of the pitcher plants (below right, with Richard Miller working on sculpts based on Ryan Church's designs) on a 4 x 4-foot table-top miniature (opposite, with Danny Wagner and Alex Jaeger). Carl Miller took stills of it, and Alex assembled a best-of collage, which we all agreed was the look we wanted. From that point on, the CG model would be compared to the collage, so the artists used it and the miniature for reference.

## Cato Neimoidia

Another cool alien planet is the homeworld to Neimoidians Nute Gunray (Silas Carson) and all the Trade Federation bad guys. Their planet consists of big stone arches that emerge from the clouds; their cities are built within hammocklike arches (top right, untextured CG hammock; bottom right, final frame). It's here that Jedi Master Plo Koon (Matt Sloan) is blown out of the sky (below, final frame).

Matt was shot in the Episode III Jedi starfighter cockpit in Sydney (opposite; later it was decided that Plo Koon should be in an Episode II Jedi starfighter, so he's actually in the wrong cockpit here)—the only real element in this sequence of shots. Everything else was a combination of CG and digi-matte paintings.

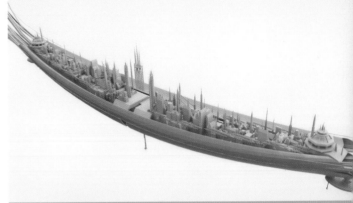

## Saleucami

As I write this paragraph in late March 2005, the Saleucami environment is active. We're finishing final shots of this planet where—following Palpatine's Order 66—another Jedi is betrayed: Stass Allie is killed while riding a speeder bike.

    Very late in postproduction, we recorded Nina Fallon, one of our coordinators, on a blue structure (opposite: I'm debriefing Nina on the finer points of speeder bike riding). This planet's environment was the result of a convoluted design process. "Saleucami" began conceptually as "bridge world," which became "Cato Neimoidia." When the scenes on Felucia were cut down, George took a part of Felucia that we'd never built—a clearing beyond the jungle where Aayla Secura is killed—which had these weird bulbous plants, and that environment was rechristened "Saleucami." So we built the bulbous plants and a couple of other flora types in CG (top right) and distributed them over a terrain that was also CG (below, final frames).

# 328

## Mygeeto

What began conceptually as Dust World was combined with Crystal World to become—Mygeeto. This is where Jedi Master Ki-Adi-Mundi (opposite, in a final frame) meets his demise following Order 66. Originally, he climbed on top of a three-legged droid and killed it, before being betrayed. For this short but complex sequence (below, final frames), George worked for a long time on detailed animatics before turning the sequence over to ILM.

## Jedi Landing Platform

Not only are Jedi massacred on far-flung planets, they are also murdered on Coruscant in the Jedi Temple itself. Senator Bail Organa, a friend of the Jedis, lands on one of the Temple platforms to investigate—when a Padawan is gunned down before him by clones. The platform was a finished set piece (opposite). The view looking past the clone troopers is the same CG Jedi Temple hangar where Anakin talks to Mace (see spread 314), but the smoke camouflages that fact (bottom right, final frame). The background behind Organa is a digi-matte painting (below, final frame).

## Bail's Airspeeder

The production art department built only half of Bail's airspeeder (bottom right, with Jimmy Smits, recorded against greenscreen and on a gimbal so it could bank and pitch). Production designer Gavin Bocquet still had the big fiberglass hull from Anakin's Episode II airspeeder, so they started from a casting of that yellow vehicle (opposite, painted and dressed). I encouraged Gavin not to worry about building the whole thing. I needed the set piece only for dialogue shots in which the principal is framed close up. For any wider shots in which the airspeeder is flying along, I knew it'd be easier to replace the whole body of the vehicle with computer graphics in postproduction. That way, we'd be able to create realistic moving reflections and other flying effects (top right, for example, a final frame). If you shoot a stationary set as if it were flying, the problem is the highlights don't move.

## Alderaan Starcruiser Hallway

This hallway was really fun to create, because it's a duplicate of the Episode IV set. We built exactly the same amount, too—an L-shaped corridor—one long straight section, which turns at a forty-five-degree angle and goes another twenty feet or so (below, with Lucas, crew, and script supervisor Jayne-Ann Tenggren in foreground). Pretty much all the Episode III action takes place in the straight part of the hallway, but there is one shot in which we pan with Bail Organa as he turns the corner.

The bluescreen in the doorway is interesting (opposite)—we never see beyond that door in *A New Hope*; it's the one the stormtroopers come through. In *Revenge of the Sith*, we find out that the ship's hangar is on the other side. When Obi-Wan arrives, he comes through that door, and you can see General Grievous's starfighter, which he's commandeered, in the background (right, final frame).

## Alderaan Starcruiser Cockpit

Another part of the starcruiser, aka the Rebel blockade runner, which we never see in *A New Hope*, is the cockpit (opposite and bottom right, built by art department supervisor Peter Russell and his crew). It's here that Yoda, Bail Organa, and Obi-Wan Kenobi discuss their return trip to Coruscant, in a scene added for the Shepperton Studios pickups (top right, final frame with a digi-matte painting of Coruscant seen out the windows, and animated graphics on the displays). The starcruiser cockpit was supposed to be reminiscent of the Republic-cruiser cockpit seen in Episode I, but really the front of the blockade runner is much wider; technically this cockpit is too small for the ship.

# 333

## Jedi Circuit Room

After landing on Coruscant, the Jedi make their way through the still-smoldering Jedi Temple to the circuit room; it's here that Obi-Wan reprograms the beacon so that it warns any remaining Jedi Knights to stay away from Coruscant.

The circuit room was originally conceived of by concept design supervisor Ryan Church, and went through various subsequent iterations. Finally, we were allowed to make a miniature, which was expensive but worth it (below, final frame; opposite: the model was made out of Plexiglas, which was laser cut and polished; the Obi-Wan cutout is there for reference, so they can line up the shot correctly against bluescreen).

## Jedi Hologram Room

Yoda and Obi-Wan Kenobi go to the hologram room, which was a half-built set in Sydney, Australia, with bluescreen at one end (opposite and below). When we originally shot this, Gavin Bocquet told me that the other half of the room was meant to be a duplicate of the existing half. So I shot a whole bunch of reference images with the intent of completing the set with a Zenviro version. But in the end, George decided the second half shouldn't be identical, and that a battle took place in this part of the room—so it became the shot-up side. The consoles were arranged a little differently, and down the hallway, replacing the bluescreen, is a view of the Jedi Temple, post-battle (right, final frame).

## Chancellor's Holding Office

For this set, where Yoda and the Emperor face off, we had a floor that went all the way from the wall to the mobile tulip (opposite, with part of the tulip—the Senate pod that rises into the Senate chamber—on the far right; bottom right, the set with tulip in center). The whole of this set was a forty-five-degree section of the oval office. I was told that the rest of the room simply continues around, so Zenviro was the perfect tool for completing the missing parts. I took reference photos, we built relatively simple geometry in postproduction, and stuck those images onto it to fill in the office (top right, a final frame—note that this particular "final" frame appears differently in the movie, after George, for technical reasons, decided to flop this shot).

## Senate Chamber

Although there are thousands of pods in the Senate, we built only two: one for Padmé (opposite, on left; middle right, final frame) and one for the Chancellor (opposite, on right; bottom right, final frame). The Episode I pods hadn't been saved; they were rebuilt in Australia for Episode II and then stored, so we saved that cost on Episode III. We also constructed many CG pods, which we needed for the big Yoda/Sidious showdown (below, final frames).

# 337

## Coruscant Skyline

A cyclorama by Yanick Dusseault, created for the sky seen outside the Jedi Council Chamber while Anakin is waiting for Mace to return.

# Mustafar

The climax of *Revenge of the Sith* takes place on Mustafar. Much of this lava planet was created with a complex, giant miniature—about twenty-by-thirty feet. ILM crew shot on this model for months (right, with lead model-maker Nick d'Abo). The lava was created with methocil (opposite), which is a food additive (the ingredient that makes milkshakes thick), mixed with kitty litter, which forms into good-looking clumps in the lava streams. For the most part, shots of the volcano landscapes started with a plate of the model and were completed with digi-matte paintings for the background. In some places, we used photos taken by master cinematographer Ron Fricke of Sicily's Mount Etna as it erupted (below, final frame).

# 339

## Mustafar Lavascape

A digi-matte painting by Brett Northcutt (opposite; below, final frame).

## 340

## Skiff Landing Platform

When Padmé decides to go in search of Anakin, she boards her Naboo skiff on Coruscant (bottom right, the set in Sydney, Australia; middle right, the final frame with the CG ship)—unbeknownst to her, Obi-Wan boards it, too, just as she flies off (opposite, CG model by Chris Stoski and Richard Bluff of the buildings for that shot).

Late in postproduction, George wanted to add technicians preparing the craft for take-off. Therefore we shot several ILM extras as if they were checking the engines, accessing a control panel. The ground crew was shot specifically to go in certain areas because they were going to be part of a big camera move (below, as planned in our layout movie). We thought we'd pulled it off, but George pointed out that two of the guys would be able to see Obi-Wan hiding, so we had to move them (top right, final frame).

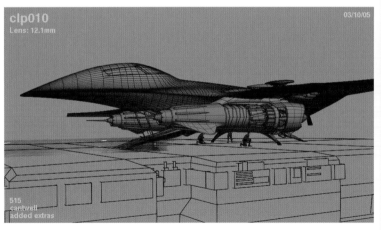

clp010
Lens: 12.1mm
03/10/05

515
cantwell
added extras

## 341

## Padmé's Starskiff Cockpit

In Sydney, Australia, George decided that he wanted to add a scene of Padmé and C-3PO (Anthony Daniels) on their way to Mustafar. There was therefore no set except green boxes, a couple of chairs, and a greenscreen. We decided to match up to similar angles of the cockpit in Padmé's Episode II ship, and that worked (bottom, on set; below, final frame). However, because the cockpit in *Attack of the Clones* belonged to a bigger ship than this one, we reduced it somewhat; there was some tricky work done on the window struts, because their number varied from cockpit to cockpit (opposite, final frame).

## Mustafar Landing Platform

Anakin arrives on Mustafar in his Jedi starfighter, so that full-sized model was placed on a patterned floor (opposite and bottom right). It's also here that Darth Vader's duel begins with his former Master, Obi-Wan Kenobi. The environment was complemented by a CG model, while surrounding rocks and lava were miniatures, but I loved the floor. Gavin Bocquet's art department used a really cool treatment for the platform that made it look like it had gone through some sort of acid bath. Gavin said it had something to do with two materials that behaved like oil and water: one layer was put down first, and then the other, but the two didn't mix; the crew blew compressed air on the platform, making the materials run, and then they let the floor dry. The result was a desolate place, which looked as if sulfur had rained on it for years (top right, final frame).

# Mustafar Control Room

Lots of violence takes place on Mustafar, and a lot of that takes place, appropriately, in the war room (also known as the control room). It's here that the Separatists gather before Vader's arrival (below, final frames). This locale was a three-walled set (opposite), whose missing fourth wall was originally going to have some steps that went down to a control-console annex. Ultimately, visual effects supervisor Roger Guyett's team simplified the missing half by just duplicating the opposite wall. The set's three doors lead to the landing platform (opposite, center bluescreen), the conference room (opposite, off camera on right), and the balcony (opposite, off camera on left).

# **Mustafar Conference Room**

Stage-space limitations made it impossible to connect the conference room to the control room, though in the movie they're linked by a doorway. Bluescreen was placed in the connecting area, and a few shots tie together visually the conference room, which was on Stage 2 (opposite), to the control room, which was on Stage 7 (bottom, final frame). To complete the set, we did put three more windows on a fourth wall and added the ceiling, which we had to show during the Jedi/Sith combat (below, final frame).

## Mustafar Balcony

One of the doors of the control room leads to this balcony (opposite, at Fox Studios; below, final frames), and from here the duel moves onto a pipe—each section built on a different stage.

## Mustafar Collection Arm

Although never stated implicitly, the planet Mustafar is being mined for energy by its inhabitants. Part of that process includes huge mechanical arms—and Obi-Wan and Anakin end up trying to kill each other on one of these collection arms. Probably the most elaborate set piece of the fight (top right, on set; bottom right, final frame of one end of the collection arm), it was built in Sydney, Australia (opposite; below left, final frame of the other end)—and only much later did we have a problem with it. As the fight evolved in editorial, George decided that the collection arm would go over a lava fall—and Anakin would have to jump off at the last second. The catch was that Anakin, if he ran straight, would run into one of the arm's "teeth" (opposite, the teeth). We talked George into not changing the design. Instead, Hayden Christensen, during the one morning of pickups shot at Elstree Studios in London, on Monday, January 31, 2005, ran down a bit of bluescreen at an angle—the last *Star Wars* live-action shot ever recorded.

## Mustafar Floating Platform

The last part of the Vader/Obi-Wan fight, which traverses (like many other *Star Wars* chases) very specific environments, takes place on platforms floating over the lava river (protected by force fields from the extreme heat). In Sydney, Australia, Ewan McGregor was put on a spring-mounted platform for wobble, and Hayden Christensen spent a few hours on a circular track (opposite, with George: Ewan is standing on a green box positioned for his jump onto the platform). Because of added lines, we shot some more of this scene at Shepperton Studios in England. Even though the actors were recorded in close proximity to each other, thanks to the digital medium, we were able to cut them out in postproduction, and move them further apart or closer together as George directed (below and right, final frames).

## 348

## Lava Riverbank

Obi-Wan flips from his platform to the riverbank, and Anakin tries to follow—but Obi-Wan at that moment slices off his former apprentice's remaining limbs. Anakin falls down the riverbank and bursts into flames. This scene was first shot in Stage 1 at Fox Studios, in Sydney, Australia, on an inclined, rectangular set dressed with rocks and dirt (opposite) where Ewan McGregor and Hayden Christensen spoke all their dialogue. Later, at ILM, a riverbank model was built for the close extension of the set (below and right, final frames). We also recorded additional footage at Shepperton Studios, with Ewan and Hayden on an even smaller rectangle of dirt, which was then positioned into a plate featuring the set from Sydney and the ILM model.

## 349

## Padmé's Starskiff Interior

Besides the Alderaan-starship cockpit, the one other complete set built for the pickups at Shepperton Studios was the starskiff interior. I traded e-mail with art department supervisor Peter Russell before the shoot, and we discussed how much set to build. I encouraged him to create the whole thing, because there was no good place to make a "split" (the separation between practical and CG set). George agreed, and therefore the scene in which Obi-Wan checks on Padmé's condition was complete in camera. Another pickup scene showed Obi-Wan stowing away in a closet for his clandestine trip to Mustafar (opposite, on set, with closet on left). One last shot, of Padmé on her medbed asking if Anakin is all right, was filmed against bluescreen at Elstree Studios and comped into a plate of the Shepperton set (bottom right, on set).

The interior was also combined with what we'd built of Padmé's ship in Sydney, Australia: a landing ramp (middle right, on set). We built a Zenviro version of the interior and used it with a plate of Obi-Wan standing atop the ramp; it's impossible to tell that the two sets were built months apart in different areas of the world (top right, final frame).

## Palpatine's Imperial Shuttle

Originally the Emperor walked down the ramp set (below, at Fox Studios in Sydney, Australia) after landing on Mustafar in his Imperial shuttle. That was cut out, but we still see the ramp in one shot: on Coruscant, after the Emperor lands at the rehabilitation center with what's left of Anakin. We could position the ramp anywhere, after all, while the shuttle and clone troopers are CG (opposite, final frame).

## Imperial Rehabilitation Center

The day we shot the return of mechanical Darth Vader, the soundstage was packed! It was a big attraction, and more crew were waiting outside the stage doors for a glimpse, which Rick McCallum had promised them. Because Hayden Christensen couldn't walk very far in the Vader suit, a temporary dressing room was constructed near the set—which was simply a portion of floor and an operating table (below). The rest of the set was completed in the digi-matte department, while the droids were CG (opposite, a global illumination render by Jeremy Cook of the rehabilitation center; top right, a final frame).

# 352

## Polis Massa Distant Crater

Obi-Wan takes the injured Padmé to a large asteroid called Polis Massa, where an archaeological team has a medical center. For the exterior, we built two miniatures: the asteroid surface, which had some curvature in the crater (opposite, with model-maker Danny Wagner); and another miniature of the inside of the crater. Since they weren't going to be extensively featured, we decided to cut some costs, and the asteroids seen floating above Polis Massa are therefore recycled ones from the Jango Fett/Obi-Wan Kenobi dogfight in Episode II. I just made them grayer this time to differentiate them from those above Geonosis, which were more brownish (below, final frame with asteroids and distant crater model).

# 353

## Polis Massa Near Crater

At ILM, visual effects DP Carl Miller lines up a shot of the more detailed crater (opposite, with model-maker Lorne Peterson on right; below, final frame completed with a digi-matte painting by Joshua Ong).

## Yoda's Observation Dome

In the Polis Massa observation dome, Yoda takes a moment to meditate. Originally he was going to hear the voice of Qui-Gon Jinn, but that idea had to be scrapped late in postproduction. At the end of the scene, Bail Organa appears in the stairwell (right, final frame). We decided to shoot Jimmy Smits as a bluescreen element in Sydney, Australia, and build the set as a miniature (below left, a study made out of foam core). A ¼-scale model was a comfortable size to work with: big enough for convincing detail, and to light and photograph effectively (opposite, the ¼-scale model with ILM's Dennis Rogers adjusting the camera; below right, final frame with CG Yoda and animated display screens).

## Polis Massa Medical Center

I thought this was a very cool set, with an interesting architectural design, which was built in Sydney, Australia (opposite, looking into the operating room; note the table is on legs). Concept design supervisor Ryan Church had chosen the color scheme because his research had revealed that these blues and greens were soothing. The operating room was meant to be glassed in, but that was added with CG in postproduction, probably because the curved glass was cost prohibitive (bottom right, looking out from the operating room to the corridor, without the glass). We also extended the ceiling and the corridor with computer graphics, painted the legs out from underneath the table—which is supposed to be hovering—and created the CG medical droids. The newborn babies were recorded against bluescreen during a separate shoot at ILM (top right, final frame).

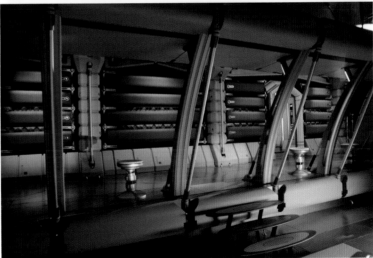

## Bail's Conference Room

Another part of the Alderaan starcruiser we hadn't seen before is its meeting room. The design follows the one established for the ship's hallway, with a curved upper wall, similar panel work, and the requisite thousands of unlabeled buttons (below, with stand-ins for the actors and a puppet Yoda for eyelines). The graphics on the wall are part of two hologram tables (opposite, the larger hologram projector is immediately below the wall graphics; the smaller one is built into the table). Apart from Yoda, this scene was complete in camera.

During shooting, Yoda's chair was operated with a string pulled by a crew member. In the final movie, however, I have to admit, since his legs don't touch the ground, I don't know how Yoda is moving his chair back and forth (top right, final frame)!

# 357

## Return to Naboo

A digi-matte painting by Yanick Dusseault and Erik Tiemens. At one point this was omitted from the movie—the painting was done but the CG Alderaan starcruiser had yet to be added. Fortunately, George put it back in.

## Padmé's Funeral

After Padmé dies of a broken heart, her body is taken back to her home planet for a state funeral. This procession was shot in many pieces: Natalie Portman was recorded in Padmé's coffin in Sydney, Australia (bottom right, on set), as was the procession accompanying her casket, though on a different day and on a different stage (opposite, with Keisha Castle-Hughes as Apailana, the Queen of Naboo). The coffin they accompanied that day was empty, as Natalie had already left the country. Governor Sio Bibble (Oliver Ford Davies) was shot during pickups at Shepperton Studios. Dozens of extras were later recorded at ILM (middle right), at various angles, so they could be multiplied into thousands of mourners (top right, final frame), while the digi-matte department filled in the backgrounds based on Erik Tiemens' concept paintings (below, final frame with digi-matte by Yanick Dusseault). The horse-creatures pulling the coffin were CG; the last element to be added was the Gungan leader Boss Nass from Episode I.

## Star Destroyer Bridge

Another replica from the original *Star Wars* trilogy. The Sydney, Australia, version had pretty much exactly the same amount of set that was built for *The Empire Strikes Back*, though that one had an attached corridor. For *Return of the Jedi*, they built the pit on either side of the catwalk, but we went back to the minimalist approach, since George needed only a couple of angles (opposite and bottom right, the set). For the one shot that required both sides, as Darth Vader walks to the window, we picked a different frame range of the recorded pit officers; we then wound it down ten seconds and flipped everybody over. We moved the one officer who was standing and pushing a button farther down the pit away from camera. The missing walls and all the other elements were completed using CG (top right, final frame).

## Alderaan

Alderaan was mentioned in George's first draft of *The Star Wars*, and is blown up in Episode IV—but in *Revenge of the Sith*, we get our first glimpse of the planet. The establishing shot of the approaching cruiser is a combination digi-matte painting with some plates shot in Switzerland by Rick McCallum and a helicopter crew (opposite, final frame with digi-matte by Brett Northcutt painted over CG building models by Jeremy Cook; below, final frame).

# Alderaan Queen's Balcony

Just a tiny bit of Alderaan was built. When Bail gives the infant Leia to his wife, the Queen, all we had was a balcony bench and some floor (opposite); the rest was CG (right, final frame). In Sydney, Australia, Jimmy Smits handed off a rubber baby; for the insert close-up, we recorded editor Roger Barton's baby, Aiden (who was also used for the infant Luke's close-up).

## Tatooine Homestead

This environment was assembled primarily in the digi-matte department from photographs taken by some of the painters (opposite, final frame). The "igloo" and berm were re-created largely thanks to stills I took during the Episode II location shoot. Ewan McGregor was filmed giving Luke (another rubber baby) to Beru (Bonnie Piesse) in Sydney, Australia, where the only set was a green ramp (below, with a stand-in). Joel Edgerton, who plays Owen Lars, was recorded much later against bluescreen, standing on another wooden ramp to simulate the edge of the crater.

■ 00004552

## 363

## Credits

It used to be that film credits listed the department heads and the cast, and that was it. But a transition occurred in the 1970s, when everyone started to be listed. George's first picture, *THX 1138* (1971), has the old-style credits, but his subsequent pictures have the new-style credits. Now pretty much everyone admits, the credits are way out of control. At ILM, we've been told to prioritize, so we list only the 500 people who have done the most hours. Consequently, while we were debating a decision on how long the end credits were going to run for one of the prequels, I asked George, "Are we really going to do a seventh reel just for the credits? Who is responsible for this?"

And he replied, "Well the story is this: In the early 1970s there was a filmmaker, who was kind of young and naive and idealistic, and who had this idea that he's doing a musical and he didn't want the film to finish with just 'The End' and some credits run in silence. He thought it'd be a good idea to have some music for the end of the movie, so when people are getting up and walking out of the theatre, they're not just walking out to silence; they're going out humming a happy tune and feeling good. It also seemed like an opportunity, since he didn't have much of a budget, to thank a lot of people who didn't get compensated very well. So he listed everybody—everybody! The studio execs were totally opposed to that and told him, 'If you do this, you're going to ruin the film industry!' But this guy didn't listen to them, and did it anyway, and it actually did sort of ruin the industry. And that filmmaker was me and the movie was *American Graffiti* [in 1973]."

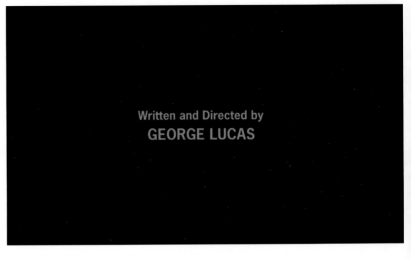

Produced by
**RICK McCALLUM**

## Archives 1

A panorama of the interior of one of Lucasfilm's archive buidings, which includes models and props from the *Star Wars* films and many other Lucasfilm projects (including Darth Vader's flagship the *Executer*, on right; below, a final frame from *The Empire Strikes Back*).

## Archives 2

A panorama of another archive, which is used to house the overflow of costumes, miniatures, models, and props (including Gasgano's Podracer, on the right; below, final frame from *The Phantom Menace*).

*This book is dedicated to my parents Glenn and Gladys Knoll,
who encouraged me to follow my passion.*

## Acknowledgments

This book began as a way to show some of the panoramas I had made for Episodes II and III, along with some commentary about what techniques we used to complete the environment, and what the final result was. This simple idea expanded greatly into coverage of all six *Star Wars* films. This meant large amounts of research to gather images from the first trilogy to illustrate the points I was trying to show. I had great support from Scott Carter and Tina Mills at Lucasfilm Image Archives, who let me spend about sixty hours pawing through thousands of historical stills to find just the right ones to use here. (Right: A rarely seen Darth Vader costume design concept from 1976.)

Jonathan Rinzler always had words of encouragement whenever I was feeling overwhelmed and pessimistic about being able to finish on time. As I got busier and busier during the final months of postproduction on Episode III, he did a huge amount of work to help get the book done.

Lorne Peterson loaned me his personal photo album from *A New Hope*, and scans from some of those stills appear in this book. Jon Erland similarly provided me with a couple of great pictures. Paul Huston, Steve Gawley, and Dennis Muren helped explain what I was seeing in some of the images.

Yusei Uesugi, Jeremy Cook, Richard Bluff, Yanick Dusseault, and John Helms all helped scrape together the really cool global illumination images in the Episode III section.

Jason Snell took a large number of reference stills while we were shooting on Episodes II and III, and a number of those were useful here.

George Lucas was very patient with me asking a thousand questions about what happened during this or that shoot, or why he chose to do something a particular way. During the last five months of postproduction on Episode III, every time we met I asked him at least one or two book-related questions.

Thank you to my wife Jennifer and my children Sarah, Lisa, Alexander, and Jane, who all offered support and encouragement when I needed it the most.

738

# Index

LIBRARY OF CONGRESS CATALOGING-IN-PUBLICATION DATA

Knoll, John.
   Creating the worlds of Star Wars: 365 days / John Knoll with J. W. Rinzler.
     p. cm.
   Includes bibliographical references and index.
   ISBN 0-8109-5936-4 (hardcover: alk. paper) 1. Star wars films—Miscellanea.
   I. Rinzler, J. W. II. Title.

   PN1995.9.S695K66 2005
   791.43'75—dc22
              2005014122

Published in 2005 by Harry N. Abrams, Incorporated, New York. All rights reserved.
No part of the contents may be reproduced without written permission of the publisher.

Printed and bound in China
10 9 8 7 6 5 4 3 2 1

Harry N. Abrams, Inc.
100 Fifth Avenue
New York, N.Y. 10011
www.abramsbooks.com

Abrams is a subsidiary of

LA MARTINIÈRE
GROUPE

**Lucasfilm Ltd.**
SENIOR EDITOR  Jonathan W. Rinzler
ART DIRECTOR  Iain R. Morris

**Abrams**
PROJECT MANAGER  Charles Kochman
DESIGNER  Scott A. Erwert
SENIOR PRODUCTION MANAGER  Maria Pia Gramaglia

*Lucasfilm would like to thank:* Scott Carter, Leland Chee, Amy Gary, Michelle Jouan,
Tina Mills, Dennis Muren, Ellen Pasternack, Sue Rostoni, and Howard Roffman.

*Abrams would like to thank:* Jon Cipriaso, Shawn Dahl, Eric Himmel, Michael Jacobs,
Dan Johnson, Isa Loundon, Samara Klein, Mark LaRiviere, Brady McNamara, Jutta Pakenis,
Stanley Redfern, and Harriet Whelchel.

**Photograph Credits**
Sue Adler, Sean Casey, Douglas Dawson, Jon Erland, Jonathan Fisher, Greg Gawlowski,
Dan Gregoire, Keith Hamshere, Tom Hilton, John Knoll, Paul Huston, Alex Ivanov,
Giles Keyte, Derryck Menere, Carl Miller, Tina Mills, Ralph Nelson Jr., David Owen,
Lorne Peterson, Jason Snell, Paul Tiller, Lisa Tomasetti, Giles Westley, and George Whitear.

*CD-ROM requirements:* QuickTime, Flash Player 6
*Keyboard directions:* Control (zoom in); Shift (zoom out)
(Note: Users may experience better performance if they copy the CD-ROM files to their
hard drives.)

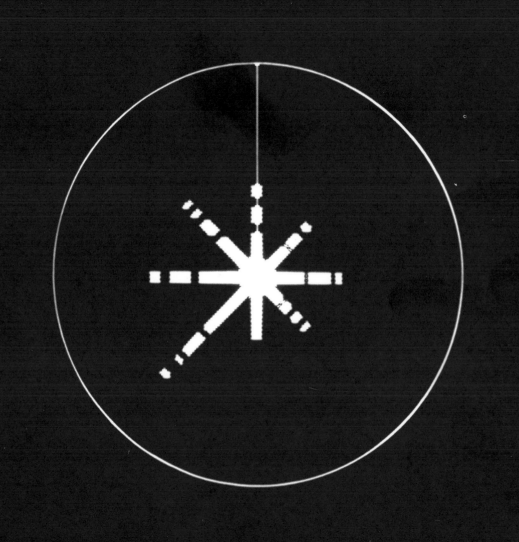